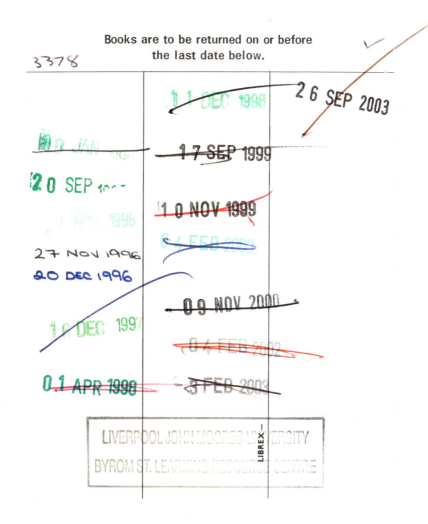

Two hundred years of British farm livestock

The Pass of Leny: Highland cattle going to Falkirk Tryst.
Gourlay Steell painted this picture around 1870 (from the Royal Collection).

Two hundred years of
British farm livestock

Stephen J G Hall

&

Juliet Clutton-Brock

BRITISH MUSEUM (NATURAL HISTORY)

Foreword by

HRH THE PRINCE OF WALES

British Museum (Natural History)

Book designed Gillian Greenwood
Ⓒ Bristish Museum (Natural History) 1989, 1991
First published 1989
Reprinted 1991
Published by the British Museum (Natural History)
Cromwell Road, London SW7 5BD

British Library Cataloguing in Publication Data
Hall, Stephen J.G.
Two Hundred years of British Farm Livestock
1. Great Britain.   Livestock.   Rare breeds,
to 1989
I. Title   II. Clutton-Brock, Juliet
636.08'21'0941

ISBN 0-565-01077-8

Typeset by Keyspools Ltd., Golborne, Lancs
Printed by Craft Print, Singapore

Contents

KENSINGTON PALACE

Britain's native breeds of livestock have had an enormous influence on the farming systems of the modern world. Today, they represent a priceless heritage of genetic variation. This variation will, for the animal breeders and farmers of the future, provide the means to adapt farm animals to new requirements. We must always be mindful of the need to protect our livestock heritage and it was to further the aims of genetic conservation that I agreed to become Patron of the Rare Breeds Survival Trust.

For many of us, the diversity of our cattle, sheep, pigs, goats and farm horses is one of the most attractive features of Britain's countryside. This diversity arose through the adaptation of livestock to local conditions and to commercial needs. I may be wrong, but I have a feeling that the wheel will turn full circle and the traditional British breeds, now rare, will in due course return to favour with an increasing number of farmers. For instance, there is already a growing demand for quality British beef on the continent, as opposed to beef from the huge continental animals. Likewise, cattle breeders are beginning to realise the disadvantages and dangers that stem from crossing our traditional breeds with continental animals and the calving problems which are inflicted on our smaller-framed animals. It is fitting, I think, that this book should have been planned to coincide with '1989 – A Celebration of British Food and Farming'.

The pictures contained in it also show how farm animals have provided different forms of inspiration to artists. They have breathed life into a landscape, or have provided the opportunity to explore textures, the play of light and shadow, or the arrangement of muscles in movement and in repose.

This book celebrates the richness of our livestock heritage and in doing so it contributes to the safeguarding of these breeds, for the benefit and enjoyment of future generations.

Charles

HRH The Prince of Wales

Acknowledgements

The inspiration for this book came from discussions with Michael Rosenberg CBE about possible contributions to the 1989 Celebration Year of British Food and Farming. We are greatly indebted to him and to the Rare Breeds Survival Trust for their support.

The names of all those who have so kindly supplied us with reproductions of pictures and photographs are listed separately but we should like to state here how grateful we are to them for all the time and trouble they took with our requests.

For the production of the book at the British Museum (Natural History) we are very greatly indebted to Susannah Farley-Green, Isobel Smith, and Chris Owen of the Publications Section and John Downs of the Photographic Unit for their patience and rigorous attention to detail in the picture research and editing.

Our thanks are due to Lawrence Alderson (Rare Breeds Survival Trust), Dr J.G. Hall and Mr J.J. Baldwin, all of whom made many helpful comments after reading the manuscript.

Stephen Hall is indebted to Professor P.A. Jewell for the hospitality of his laboratory in the University of Cambridge.

Preface

The impetus for this book has been the 1989 Celebration Year of British Food and Farming which itself celebrates the 150th anniversary of the Royal Agricultural Society of England and the centenary of the Board of Agriculture (the precursor of the Ministry of Agriculture, Fisheries and Food). Published by the British Museum (Natural History) under the aegis of the Rare Breeds Survival Trust and with the invaluable support of Michael M. Rosenberg, CBE, this book is dedicated to the celebration year and to the greater understanding of this land, its history, and its resources.

Diane Rosher '86

1   Introduction

The second half of the eighteenth century saw a gathering of momentum in industrial and agricultural change that was unprecedented in world history. In the British Isles capital flowed into industry from the land and back again for the increased production of food. At the beginning of the century the population had been estimated at $5\frac{1}{2}$ million; by 1801 it had reached $10\frac{3}{4}$ million. Malthus watched this increase in the numbers of people and wrote his famous *Essay on The Principle of Population*, published in 1798. The great controversy aroused by this work has been reviewed by Blaxter (1986). Malthus maintained that the number of people supported by a given resource of land could only be increased by a constant number each year, while populations could grow like money invested at compound interest, that is, with an ever-increasing number of people added each year. He wrote:

> The food therefore which before supported seven millions, must now be divided among seven millions and a half or eight millions. The poor consequently must live much worse, and many of them be reduced to severe distress. The number of labourers also being above the proportion of the work in the market, the price of labour must tend toward a decrease; while the price of provisions would at the same time tend to rise. The labourer therefore must work harder to earn the same as he did before. During this season of distress, the discouragements to marriage, and the difficulty of rearing a family are so great, that population is at a stand. In the mean time the cheapness of labour, the plenty of labourers, and the necessity of an increased industry among them, encourage cultivators to employ more labour upon the land; to turn up fresh soil, and to manure and improve more completely what is already in tillage; till ultimately the means of subsistence become in the same proportion to the population at the period from which we set out. The situation of the labourer being then again tolerably comfortable, the restraints to population are in some degree loosened.

Malthus, like many others of his day, was aware of the desperate need for changes in the methods of food production for a country where the population was increasing at a great rate. Today it can be seen just how successful these changes were; Britain alone has a population of 55 million and there is a surplus of meat and grain throughout the western world.

By the second half of the eighteenth century every farmer and landowner was concerned with the need for an increase in food production. This had begun much earlier with the enclosure of the land into small fields; hundreds of thousands of hectares of wasteland and woodland had been drained and felled, and roads had been much improved by as early as 1730. During the reign of George III ('Farmer George', 1760–1820) private Acts of Parliament were passed to overrule individuals who were resistant to enclosure. New methods of agriculture were introduced, notably the growing of root crops such as the turnip which enabled many more livestock to be fed throughout the winter. The most famous early improvers were 'Turnip Townshend' (1674–1738), who popularized crop rotations, and Jethro Tull (1674–1741), pioneer of farm mechanization.

Agricultural societies were set up: the Bath and West in 1777, the Highland in 1822 (its forerunner, the Honourable the Society of Improvers in the Knowledge of Agriculture in Scotland was founded in 1723), and in May 1838 a Society was founded in London which soon after its first field meeting at Oxford acquired its Royal Charter and became known as the Royal Agricultural Society of England. This significant meeting at Oxford of 1839 was the first Royal Show.

At the same time the great names in livestock improvement were at work, having learned their skills from the breeders of racehorses. The first stud book for horses was founded in England in 1791 and became known as the *General Stud Book* in 1808; herd books for cattle and pigs and flock books for sheep followed. *Coates Herd Book* (for Shorthorn cattle) was founded in 1822 and the Shorthorn Society was incorporated in 1875.

It is well known that Henry VIII, in 1535, passed a law stating that no stallion under 15 hands (152 cm) and no mare under 13 hands (132 cm) was to be kept alive. He also imported many horses from Italy and Spain for his Royal Stud at Hampton Court. So began a long line of interest by royalty and nobility in horse breeding and imports culminating in the three Arabs, Byerley Turk (in 1689), Darley Arabian (c. 1700), and Godolphin Arabian (in 1730). By the time Robert Bakewell began his experiments in improving sheep and cattle these Arabs had already changed the conformation of the British horse with the creation of the Thoroughbred. Eclipse, the most famous of all racehorses, was descended from both the Darley Arabian and the Godolphin Arabian, and he died 200 years ago, in 1789. He was the sire of 335 winners, and his life spanned the time of greatest advancement of livestock improvement, from 1764 to 1789.

Although rather little is known about the life of Robert Bakewell he is generally regarded as the apostle of livestock improvement. He lived from 1725 to 1795 and succeeded his father in the tenancy of Dishley Grange, near Loughborough in Leicestershire in 1760 when he was thirty-five years old, after he had already begun his improvements on the Leicester breed of sheep and the Longhorn breed of cattle. His methods were based on the principle of inbreeding which was already proving to be so successful with the racehorse.

The results of Bakewell's work led to the reshaping of British livestock and to the inspiration of all subsequent generations of animal breeders. Furthermore it was from these breeders that Charles Darwin first developed his ideas on natural selection as being analogous to artificial selection and, in turn, once the processes of change had been explained the breeders were

Eclipse, painted by Francis Sartorius (1734–1804). The most influential racehorse of the eighteenth century, Eclipse was foaled at Windsor during the solar eclipse on 1 April 1764. Early racehorse breeders showed the way for livestock farmers to improve local breeds, adapting them to meet the demands made by a rapidly increasing human population.

increasingly able to apply their practical knowledge to the moulding of farm animals for greater efficiency and production. Before Bakewell's time it was believed that the only way to obtain higher productivity was to feed animals better, and the usual practice was to send the best animals to market and retain the poorest for breeding. About the middle of the eighteenth century Bakewell and his followers began to realize that by careful selection of the progeny of favoured animals a breed could be changed dramatically. Bakewell also played a major part in the development of a new profession, that of the specialist breeder of pedigree sires for sale or hire to commercial farmers.

The necessity of feeding the growing urban populations prompted the development of meat characteristics in lowland sheep and cattle. In cattle breeding Bakewell's aim had been the creation of a profitable meat animal. While he did not achieve this with the Longhorn, his followers succeeded with the Shorthorn. Bakewell's personal success was, however, indubitable with his new breed of sheep. Here he used the long-woolled sheep of Leicestershire and Lincolnshire which in the early eighteenth century were the main source of London's mutton, fattened for the market at three to four years old when culled from the wool-producing flocks.

Robert Bakewell began his programme of sheep improvement in the 1740s and probably built on the work of an earlier breeder, Joseph Allom. Opinions have been expressed that the Ryeland, at the time a small, slow-growing, fine-woolled heath sheep, was also used, but it is almost certain that the important founder stock came from Lincolnshire. The new breed had been fixed in type by 1770, and it was in this year that Arthur Young (1741–1820), Bakewell's most articulate admirer and publicist, visited Dishley. Bakewell's pre-eminence in the history of animal breeding is largely due to Young's enthusiastic advertising of his achievements.

In unimproved sheep, as in mammals generally, the growth and development of the parts of the body takes place in the following sequence; first the skeleton and internal organs, then the muscles, and finally the fat. Modern farm animals intended for meat production have been selected to begin deposition of fat at a younger age while bone and muscle are still developing, and this gives succulence to the meat. In meat lambs the fat is deposited within the muscles rather than under the skin or around the kidneys and intestines. However if kept too long such sheep become over-fat and it was clear early on that Bakewell's New Leicesters were ready for sale a year earlier than any other breed.

This was Bakewell's major contribution to sheep breeding. While the quality of the Dishley mutton was not high and the sheep were poor breeders, crosses with other breeds and subsequent selection rectified these faults without losing early maturity. For example, the Lincoln, after crosses of Dishley rams did not have the same tendency to over-fatness as Bakewell's New Leicester, and in the early nineteenth century became more popular for meat production. Culley's emergent Border Leicesters were further selected for prolificacy, which had not been a feature of Bakewell's sheep.

Before long there were New Leicester flocks all over lowland Britain, and crosses had been attempted with most other breeds. However, the longwools were soon superseded by the Down breeds, and today while most British sheep have some New Leicester ancestry the only comparatively pure descendants of Bakewell's stock are the Border Leicester, and the Leicester Longwool.

It was not only sheep and cattle that received Bakewell's attentions; he levelled his grasslands and set up an irrigation system on his farm, improved the black heavy horses of Leicestershire, and experimented with pig breeding.

The agricultural revolution of the second half of the eighteenth century, for so it may be

Robert Bakewell (1725–1795), the acknowledged pioneer of modern livestock breeding, painted by Boultbee (1745–1812).

termed due to the great changes that took place, was radical and dramatic, and laid the foundations of today's British livestock industry. Many of the features we now regard as inseparable from animal husbandry are the results of nineteenth century developments, for example the dairy industry and veterinary science, neither of which was of much account in the eighteenth century. The Veterinary College of London was founded in 1791 (later to become the Royal Veterinary College) but until 1872 its studies were confined to the diseases of horses. The practice of veterinary medicine was improving, however, and in the 1830s and 1840s William Youatt (a veterinary surgeon who died in 1846) wrote his monographs on the dog, horse, pig, sheep, and cattle. These works are not referred to today for their medical advice, even though Youatt had an attitude of unusual sensitivity and common sense, but they are still the standard source of information on breeds of the early nineteenth century, augmented by *A history of British livestock husbandry* by R. Trow-Smith (1959).

The enclosure of common land continued until the late 1860s; there was much poverty and a huge increase in population, not, as Malthus had predicted, from an increasing birth-rate, but from a lowering of the death-rate. This was the impetus for a great exodus of people and livestock from Britain to the new colonies.

In 1793 the Prime Minister, William Pitt, established the Board of Agriculture, with Sir John Sinclair as President and Arthur Young as Secretary. Surveyors were appointed whose reports on the various counties, combined with those of William Marshall who was active before the Board was formed, provide indispensable information on the farming of the period. The Board was one of the casualties of the post-Napoleonic Wars era of change and it was dissolved in 1822, but re-established in 1889 in response to the findings of the Richmond Commission (1879–82) on agricultural distress. It was renamed the Ministry of Agriculture and Fisheries in 1919.

For the text and illustrations of this book, extensive use has been made of the works of the eighteenth and nineteenth century agriculturalists and artists. Of the latter, one of the most important was George Garrard, who by means of sculptured models (made to a scale of two and a quarter inches to the foot; just under one-fifth of full size), engravings, and records of the dimensions of individual animals, left a very clear and accurate image of the appearance of the livestock of his time. In 1800, when most of his engravings were published, Garrard wrote this introduction to his work:

> These works are not intended merely as matters of curiosity, they exhibit, at once, the ideas of the best judges of the times, respecting the most improved shape in the different kinds of Live Stock – Ideas which have seldom been obtained without great expence and the practice of many years. It is presumed that, by applying to works of this kind, the difficulty of acquiring a just knowledge upon the subject may be considerably removed; and also, that distant countries, where they may be sent, will be enabled to form very perfect ideas of the high state of cultivation in which the domestic animals are being produced at this day in Great Britain; and should further progress be made, these Models will shew what has already been done, and may be a sort of standard whereby to measure the improvements of future times.

The accuracy with which Garrard depicted his subjects was not repeated by later artists, but the fashion for producing engravings and paintings of livestock continued and in the nineteenth century was exemplified by David Low in his great two volume work on *The Breeds of the Domestic Animals of the British Islands* published in 1842. Low, Professor of Agriculture in the University of Edinburgh, was a great protagonist of improvement and he intended setting up a University Museum of Agriculture. He began in 1832 by commissioning the artist William Shiels to paint in oils all the notable breeds of British livestock; half life-size of cattle and horses, life-size of the others. Many of these paintings were redrawn, reduced, and engraved by Nicholson for publication in Low's book, and selections of them are reproduced here.*

Low understood the principles of animal breeding to an extraordinary extent, considering that he was writing sixteen years before the publication by Darwin of *The Origin of Species* in 1858. In the introduction to his book of 1842 Low makes the following comment which incorporates many of the tenets of today's Rare Breeds Survival Trust:

> Animals become gradually adapted to the conditions in which they are placed, and many breeds have accordingly become admirably suited to the natural state of the country in which they have become acclimated. Thus the West Highland breed of cattle has become suited to a humid climate and a country of mountains; the beautiful breed of North Devon to a country of lesser altitude and milder climate. In these and many cases more, an intermixture of stranger blood might destroy the characters which time had imprinted on the stock, and give as the result a progeny inferior in useful properties to either of the parent races. Not only have individual breeders erred in the application of this kind of crossing, but it is certain that several entire breeds have been lost which ought to have been carefully preserved.

Later, in 1845, David Low published another work, *On the Domesticated Animals of the British Islands*. This did not contain the illustrations of Shiels or Nicholson but was intended as a textbook for agriculturalists and it remains, like the monographs of Youatt, amongst the most informative books on livestock breeding in the nineteenth century.

The principle that has been followed in the production of the present work has been to pursue the standards of Garrard, Youatt, and Low. The pictures reproduced here, by the best artists of their day and wherever possible of named animals, are used to illustrate the history of the breeds of livestock over the last 200 years. They have been chosen to be in as many different media and styles as possible in order to provide a balanced view of a subject that has been very widely illustrated over these two centuries.

At the present day the remarkable situation appears to have been reached where agricultural production in the western world is sufficient to feed the population, and retrenchment may have to become part of agricultural policy. The promotion of the old breeds of livestock that are so well adapted to the different regions of the country may again become of economic importance as the animals of the factory farm are once more put out to grass. Perhaps even more importantly, the native breeds of livestock, whose histories are outlined in this book, should be conserved simply because they are part of our living heritage and their presence in the countryside enhances the quality of human life.

* Only a few of Shiels' original paintings survive and many of them are in poor condition. They are now held in the Royal (Dick) School of Veterinary Studies and in the National Museums of Scotland, Edinburgh. Several have recently been restored.

2 Western and northern cattle

Daniel Defoe in his tour of Britain, published 1724–6, described how great numbers of cattle from the Highlands of Scotland were driven each year to be fattened on the Norfolk marshes for the London beef market. The trade reached a peak in 1835 when maybe as many as 80 000 cattle crossed the border. Polled cattle were easier to manage than horned on the drove roads, so breeds that were naturally hornless were in favour. Of these the Galloway was always the most numerous but there were also hornless cattle in Sutherland, Ireland (the Irish Moiled), and the Hebrides. The polled cattle on Skye were made famous by the comment of Dr Johnson that, 'of their black cattle some are without horns, called by the Scots humble cows, as we call a bee an humble bee, that wants a sting' (*Journey to the Western Isles*, 1775). In Johnson's time, 'cattle' meant livestock including horses, sheep, and pigs, while 'black cattle' meant bovines, whatever their colour.

Wilson, in his renowned *Evolution of British Cattle* (1909) claimed that the polled breeds of the north had been brought to Britain before the Conquest by the Norsemen. This could be true but it is also likely that hornlessness, which occurs sporadically as a natural variation about once in every 50 000 calvings, was selected for by the cattle drovers for their own convenience and safety on the long marches south.

At the end of the eighteenth century there were three main categories of cattle in Scotland, the Kyloes or Highland, the Galloway polled, and the dairy breed of Ayrshire. In addition there were the polled Angus, the dual-purpose breed of Fifeshire (an ancient predominantly black, horned type), and the Shetland cattle (which occurred in a great variety of colours). Later, Youatt described a different race from almost every county of Scotland, but whether this reflected the facts or his own tendency to create a proliferation of breeds is debatable.

The cattle of Wales were little influenced by the progressive modern farmers of the eighteenth century. There were three main types of indigenous Welsh cattle; the white park (feral) breeds, the red breeds with white or dark markings, and a black type with or without white. The latter two were described in the old literature as middle-horned. The most notable of the old Welsh cattle were perhaps those from Anglesey. These were small, black cattle with rather long horns which turned upwards at the ends; they were traditionally used as plough oxen and said to have been much admired by Bakewell. Like the Scottish cattle those from Wales were important to the droving trade and were driven as far away as Kent and Northamptonshire. By the middle of the nineteenth century, however, drovers had vanished from almost everywhere, having been supplanted by radical changes in the pattern of meat distribution.

Included in this chapter are the breeds that are native to Scotland, Wales, and Ireland, as well as the White Park, Chillingham, and British White.

The north-east of Scotland has always had a very strong tradition of arable agriculture and it was in this context that the Aberdeen-Angus was developed. Unlike all the other British beef breeds except the Shorthorn, the Aberdeen-Angus emerged as part of a sophisticated arable husbandry, where stall feeding and liberal use of concentrates were practised. This cow, painted by an unknown artist in the early nineteenth century, is in an indoor setting, perhaps to make the point that she has been raised under carefully controlled conditions.

Aberdeen-Angus

The Aberdeen-Angus is Scotland's most famous breed of cattle and it is perhaps surprising that it is one of the most recently established. The pioneer breeder was Hugh Watson of Keillor, near Dundee, who started showing his black polled cattle in 1820. He sent some of his stock from the Highland Show at Perth in 1829 to Smithfield, where they made a very favourable impression. By the 1840s consignments of meat from Aberdeen were transported principally as jointed carcasses, transport costs rendering it most profitable to send only the most expensive cuts. The railway route to London was completed in 1850.

Against this background of a new, sophisticated trade in prime beef, Watson's herd developed. He is regarded as having fixed the type of the new breed and by the time his herd was dispersed in 1861 it had been highly selected within itself. For fifty years of its existence Watson seems never to have bought in a bull. He sold stock to William McCombie of Tillyfour, near Aberdeen, who attached the same priority to meeting the requirements of the London trade. McCombie's father had purchased Tillyfour and the neighbouring farms from the proceeds of his cattle dealing business and knew how to breed cattle that would yield good carcasses, or, in the terminology of the period, that would 'die well'.

The early breeders managed to develop carcass quality without losing milking ability, so up to 1914 the breed was represented at the London Dairy Show. Its main rival as a beef breed in Scotland was Cruickshank's Scotch Shorthorn which could be fattened more rapidly but which did not milk so well and was less hardy. The characteristics of the two breeds were combined by crossing them, and the Aberdeen-Angus cross Shorthorn became the source of most of the prime beef produced in Scotland. Aberdeen-Angus bulls were also very suitable for crossing with small hill cows of various local types.

The Polled Cattle Herd Book, which included Galloway registrations, was started in 1862, and the Aberdeen-Angus Cattle Society inaugurated in 1878. The breed was established in Australia by the 1850s and soon after in New Zealand, Canada, South America, and southern Africa. In 1891 a separate class at the Smithfield Show was provided for the breed, which has since been consistently successful there. It was the first breed to provide a champion at two years old. Up to 1955, of the 333 championship prizes awarded, 158 went to pure-bred Aberdeen-Angus cattle, and 107 to cross-breds with Aberdeen-Angus blood (Shorthorns won 40, Galloways 14, Herefords 9, Devons 4, and Welsh Black 1). This success has been continued in recent years but the prize lists now usually include Charolais × Aberdeen-Angus crosses.

The big sales for Aberdeen-Angus cattle (as well as for Beef Shorthorns) have always been the Perth bull sales, at Macdonald Fraser & Co. Between 1865 and 1977, 71 433 bulls of

An Aberdeen-Angus bull (Pirate of Monkwood, born in 1978, the property of the Scottish Milk Marketing Board) accompanies Shorthorn cows in the Market fields, Perth, overlooking the River Tay; painted in 1982 by Eirene Hunter. This picture emphasizes current breeding policy to increase size and growth rate.

Gourlay Steell RSA (1819-1894) was appointed to Queen Victoria in 1872 as animal painter for Scotland, and was also official animal portrait painter to the Highland and Agricultural Society. This portrait of the Guisachan herd of Lord Tweedmouth was painted in 1890. The herd was founded in 1878 and dispersed fifteen years later. The cattle shown here are Field Marshal of Guisachan, Frailty (a great breeding cow of her day), Cash (her son: Jubilee Champion and winner of the Queen's Gold Medal, Windsor 1889), Pride of Guisachan 20th and Fame of Guisachan.

these breeds were sold there, for a total of £10 943 515 5s 0d. The world record, still unbroken for any breed, was set at Perth in 1963 when Lindertis Evulse was sold to the USA for 60 000 guineas.

During and after World War II it became important to increase beef production rapidly, and a subsidy scheme was started to encourage dairy farmers to inseminate their best cows using the appropriate breed of dairy bull, so that herd replacements would be the daughters of good milkers. Cows that were poor milkers were put in calf to beef bulls. The scheme was in full swing by 1952–3 when artificial insemination was operating, and at this time 15 per cent of dairy inseminations were by beef bulls. The cross-bred calves were reared for beef. Before the subsidy was paid, proof had to be given that the beef bull had been the sire; colour-marking bulls were therefore used, namely Aberdeen-Angus and Hereford. The progeny of these bulls were instantly recognizable by their black colour and lack of horns (Aberdeen-Angus), and white face (Hereford). Another advantage accrues to the Aberdeen-Angus bull; he sires calves of low birth weight and is therefore a very good mate for young heifers. In recent times more and more dairy heifers have been inseminated before they are two years old.

The 1950s and early 1960s was a successful period for the Aberdeen-Angus breed but it was not to last. As with all beef breeds the late 1960s saw great changes. In 1959, the post-war peak year, there were 9676 registrations, but since then numbers have fallen steadily and 1970 was the breed's worst year when the champion at Perth sold for a mere 700 guineas. In spite of there being around 50 million Aberdeen-Angus cattle world-wide, exports of bulls have dwindled to nothing from their early 1960s peak of around 250 annually. In the north-east of Scotland, the breed's stronghold in Britain, the small hill cows which crossed so well with the Aberdeen-Angus became, since the 1950s, too expensive to keep and on the best ground barley was far more profitable than beef. The demand grew for large fast-growing animals and the Charolais in particular became the favoured bull.

The response by breeders of the Aberdeen-Angus to new competition has been to increase size, and the import of large bulls started in 1971, pioneered by the Tangier herd with two bulls from Canada. This modernization of the Aberdeen-Angus by bringing in new blood to increase size was foreshadowed eighty years ago in Australia. There the Murray Grey originated when, starting in 1905, a series of grey calves was born to a light roan Shorthorn cow which had been mated with an Aberdeen-Angus bull, on a farm in the upper Murray River valley in Victoria. The black colour of the Aberdeen-Angus is dominant to other coat colours, so this series of grey calves was surprising, and initially the calves were kept as a curiosity. In due course the female calves were crossed with Aberdeen-Angus bulls and the end result was a beef breed which is, in effect, a grey-dun Aberdeen-Angus. A breed society was formed in Australia in 1962 and in 1973 cattle were imported into Britain, with an affiliated breed society being formed to promote the breed. It is the only imported breed of cattle in Britain which is not of European origin.

Ayrshire

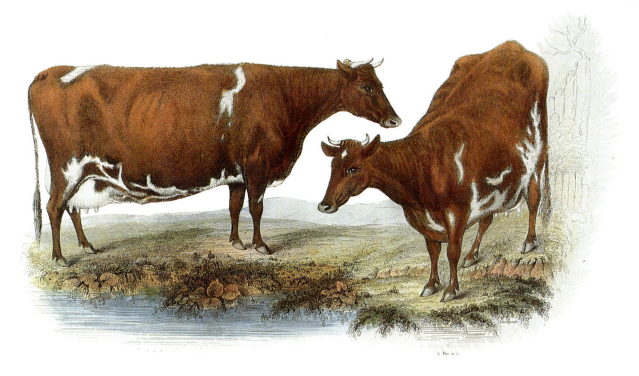

James Howe (1780–1836), from Peeblesshire in southern Scotland, painted these Ayrshire cows around 1830. He has clearly shown the neat udder characteristic of the breed. In and around Ayrshire, women did the milking and this is said to be why, from early times, small teats were favoured for these cattle. Latterly, priority was given to udder conformation, sometimes regardless of milk production. Cows of particularly good udder type were known as vessel-bred cows.

The Ayrshire arose from a cross of imported cattle with the local race which appears to have been black in colour. There are records from around 1769 of Dutch cows being brought in, conferring milking qualities and a brown, or brown and white colour, on the local stock. The new breed was usually known as the Dunlop or Cunningham. Even today, black and white Ayrshires are quite frequent, a reminder of the mixed origin of the breed.

The Ayrshire was recognized by the Highland & Agricultural Society in 1814 and the Herd Book Society was founded in 1877. By this time the breed had been exported to Canada, the USA, Finland, Australia, and New Zealand. The Swedish Government bought large numbers for thirty years from 1879. The breed did very well in its new environments and in recent years Finnish and Swedish Ayrshires have been re-imported.

It appears that when any breed of livestock becomes popular it is automatically selected for breed points rather than for its commercial features. This happened with the Ayrshire and by 1919–20 there had developed a difference of opinion on whether the show type cow or a milk producer was the desired aim. A new show standard was promoted, in which points were awarded for milk production and for udder conformation. This led to a very successful merger of show type and dairy merit which achieved great success at the London Dairy Shows of the 1920s.

'The Fifeshire breed.'
The Ayrshire is the only survivor of the medley of breeds, types and races of cattle of the Scottish lowlands in the early nineteenth century. However, the cattle of Fifeshire should be remembered; they were painted by several artists including William Shiels (whose works were engraved for Professor David Low's book of 1842). Low described Fifeshire cattle as a mixture of breeds, mainly black or black and white. He felt the only worthwhile variety among them was the Falkland, itself probably of English or Dutch origin, which by 1840 was virtually extinct. As the dairy was, in that area, merely 'an affair of the household' and financially insignificant, Low believed that Fife farmers should replace these cattle by a beef breed such as the Shorthorn.

Some of the early nineteenth century portraits of Ayrshires depict oxen and other fatstock, but the Ayrshire has been a dairy breed for most of its recorded history. The history of British cattle husbandry is usually summarized as being orientated towards beef until the end of the nineteenth century, then being dominated by the dairy industry. Thus the existence of a specialized Scottish dairy breed throughout the nineteenth century requires explanation, if only because the English equivalent, the Suffolk Dun, failed to fill a need and died out.

A vital factor in the development of the Ayrshire was the farm butter and cheese industry of the south-west of Scotland which until the beginning of the twentieth century was much more important than the liquid milk market, and this meant there was a place for the Ayrshire as a pure dairy breed. Scottish cheese-making on a wide scale dates from the late seventeenth century when a new way of making cheese from whole milk was introduced into north Ayrshire, presumably from Ireland. The Cheddar process came in during the 1850s leading to further expansion, only slowing down in the 1880s when North American cheeses began to compete. The response of dairy farmers to this was to switch to the liquid milk market which was only saved from financial collapse, due to oversupply, by the creation of the Milk Marketing Board in 1933.

It was against this background of a seller's market for butter and cheese and latterly a buyer's market for liquid milk that the Ayrshire breed developed. The difficult farming years before 1939 also, paradoxically, aided the breed. In the 1930s many Scottish dairy farmers gave up in Renfrew and Ayrshire and took tenancies in south-east England, notably Essex, moving their families, cattle, goods and chattels south, often by special train. These farmers usually succeeded where the outgoing tenants had failed and their example helped the Ayrshire breed to win many converts.

Another factor came into account after 1945; in 1948, 29 per cent of Scottish dairy herds, which were mostly Ayrshires, were tuberculin tested, compared with 7.3 per cent in England. Indeed, the 1947 herd book showed that registrations had doubled within five years, with practically every female calf registered. The quick way to establish a TT herd in England was to buy Ayrshires from Scotland, and a great many new herds were established in the south. By 1955, 18 per cent of cattle in milk producing herds in England and Wales were Ayrshires, compared with 41 per cent Friesian. Indeed it is only in recent years that the arch rival, the Friesian, has made any lasting impression in Scotland. In 1948 for example, 5713 Ayrshire bulls were licensed in Scotland compared with 389 Friesians.

Today the British Friesian is far and away the dominant dairy breed, but the Ayrshire still fills a need, particularly in Scotland, and although it has been reduced in numbers over the last few years it remains popular. The largest Ayrshire population in the world is in Finland where, in 1977, there were over half a million cows. The Ayrshire is of smaller body size than the Friesian, with a rather lower yield of high quality milk, but there are still plenty of dairy farmers who find the Ayrshire's combination of qualities is just what they need.

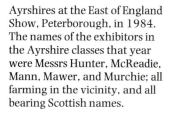

Ayrshires at the East of England Show, Peterborough, in 1984. The names of the exhibitors in the Ayrshire classes that year were Messrs Hunter, McReadie, Mann, Mawer, and Murchie; all farming in the vicinity, and all bearing Scottish names.

Galloway

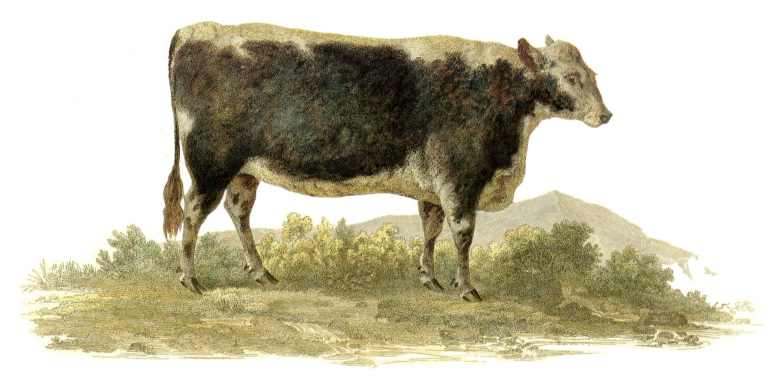

The two modern Scottish breeds of black, hornless beef cattle, the Aberdeen-Angus and the Galloway, have some superficial similarities which reflect their descent from the same primitive stock, but this descent followed very different paths. Today the former is a rapidly maturing animal responding to intensive feeding, while the Galloway is a beast for marginal and hill lands. These differences reflect the farming practices of the south-west as opposed to the north-east of Scotland.

From about 1750 south-west Scotland had been a major source of store cattle, which were taken south by the drovers to be fattened for the London market. In his *General View of the Agriculture of Galloway* of 1794, James Webster described how polled cattle were preferred to horned for breeding, and how heifers not intended for breeding were spayed at two years old. Colours were black or brown. By about 1840 the droving trade had ended. Cattle farmers in south-west Scotland turned to dairying and the beef cattle were forced to live in the hills.

Attempts had been made to improve the native beef cattle by the use of the Shorthorn but these were not successful. The husbandry of beef cattle, and with it the Galloway breed, could easily have vanished and been replaced by sheep farming, but it was discovered how a valuable cross-bred could be produced using Shorthorn bulls on pure-bred Galloway cows.

George Garrard (1760–1826) produced between 1799 and 1814 a remarkable set of engravings and models of livestock, accompanied by dimensions of the individuals depicted. The comment 'Mixture of Indian' was written under this engraving of *The Fat Galloway Heifer*. Interbreeding with imported Indian humped cattle (zebu) was tried for a short time by improving agriculturalists at the beginning of the nineteenth century, but it appears to have had no long term effect, although it may be detectable in the blood groups of a few breeds. The brindle colour, and the white back (also called finch-back or eelstripe) died out in the Galloway soon after the Herd Book was started in 1877.

Bolebec Dun Champion, bred and owned by Mr Christopher Marler, of Olney in Buckinghamshire, painted by Diane Rosher in 1985. There are several colour variants of Galloway cattle, reviewed by Lord David Stuart in 1970. The best known is the Belted Galloway in which the ground colour is usually black but in this bull it is dun.

This probably began before 1830, the year when a cross-bred of this kind was shown at the Highland Show. Now known as the Blue-Grey, the female progeny of this cross is a long lived hill cow inheriting hardiness from the dam and milking qualities via the sire. A variant of the Shorthorn, the Whitebred Shorthorn, is specialized for this cross.

It seems very likely that since the mid nineteenth century the popularity of the Blue-Grey has been the major factor in the survival of the Galloway breed; indeed by 1908 only 0.4 per cent of the national cattle herd was Galloways. Even so, the breed has always been well known to cattle farmers. For at least 200 years there have been prized herds of parkland Galloways in England, as well as in Scotland. William Cobbett, in his *Rural Rides* (1830) saw cattle of what he called the Galway-breed (obviously he meant Galloways) at Highclere, Hampshire; '. . . about forty cows, the most beautiful that ever I saw . . . their colour is a ground of white with black or red spots, these spots being from the size of a plate to that of a crown-piece . . . they were in excellent condition, and so fine a sight of the kind I never saw'. Culley also wrote about ornamental herds of Galloways as early as 1786.

The breed was also put to practical use in the development of other breeds. In the late eighteenth century Galloway blood (the 'Galloway alloy') was introduced into the Shorthorn,

rather controversially, and it is quite possible it played a part in the formation of the Red Poll.

During World War II the value of the pure-bred Galloway for hill grazing was recognized and numbers were expanded under Government encouragement. However, this form of marginal land use depends critically on the level of state subsidy and today in Britain it is not profitable, despite the well-established fact that cattle and sheep grazing together improve the pasture for each other. In West Germany, in contrast, beef production from heath, marsh, and upland is subsidized, making the Galloway a popular choice, and high prices are being paid for good stock from Britain. There are now at least 1800 Galloways in Germany.

Blue-Greys have always been popular in the north of England and in Scotland and can be seen on relatively poor grazing in many parts of England. Nowadays they are mostly used as suckler cows with Charolais or Limousin cross calves. Most such herds calve in early spring while still in winter housing, although Blue-Greys will happily winter outdoors. The cow draws on her body reserves in early lactation and the herd is turned out to pasture when the

The Galloway bull of 1830, painted by Howe, was very like that of today. The long, thick coat, short and broad face, hairy ears, and lack of a markedly domed poll distinguish the breed from the Aberdeen-Angus.

ground is dry enough and grass growth has started.

As a suckler cow the Blue-Grey invites comparison with the Hereford-Friesian. Generally the latter is more productive, but does not live so long, and Blue-Greys are usually sold at a premium. The physiology underlying this difference has been tentatively looked at and it seems that Blue-Greys have more total body fat than Hereford-Friesians of the same body condition. The latter may have to draw on muscle tissue to sustain lactation and this may be why this cross-bred does not 'wear so well'.

James Howe made 'many hundred sketches of a unique character', depicting Scottish livestock; this one, of Highland cattle being driven south, dates from around 1830.

Highland

Long-horned, shaggy-coated Highland cattle are Britain's most distinctive breed, and today as ever they are strongly evocative of the romance and mystery of the Highlands. Up to the mid nineteenth century Highland cattle were familiar all over Scotland and most of England as they were the foundation of the droving trade. As young animals they foraged around the hills and sea-lochs of the western Highlands and Islands; at four to five years old, weighing about 150 kg, they were gathered and driven to market at Crieff. They covered about ten miles (16 km) a day, swimming when faced with rivers or short sea crossings; some were fitted with shoes before or during the journey. They were then sold to specialist graziers, who fattened them to a slaughter weight of about 250 kg. This trade eventually declined because people wanted the better quality beef that could be obtained from young animals reared and fattened on the new fodder crops near to the market.

The Highland breed includes cattle of several colours. In 1984, of the 53 stock bulls listed by the Highland Cattle Society, 31 were red, 12 yellow, 4 black, 3 brindle, 2 white, and 1 silver. Belted and pied patterns are, however, not known.

The history of Highland cattle in the eighteenth century was poorly recorded, probably because the animals were so commonplace, like the 'Irish stores' of the twentieth century. However, in the nineteenth century certain lairds, notably the McNeil of Barra, the Duke of Hamilton, and the Duke of Argyll, encouraged development of the breed; the Stewart brothers of Harris were also noted improvers. Stock seems to have been selected from island and mainland populations, with no evidence of lowland blood having been brought in. Hardiness has remained a key characteristic of the breed, which is quite closely related to the Galloway.

The breed society was founded in 1884, with 561 bulls listed in the first herd book, most of which were black or dun. Some went to Canada in 1882, and in the 1920s exports were made to Wyoming (USA) and Argentina. Today they are gaining a following in West Germany and elsewhere in Europe. In Britain there are now about 179 herds or, more correctly, 'folds'. Highland cows are often crossed with a Shorthorn or similar bull to give a milky cow suitable for suckling a cross-bred beef calf.

There is no written information with Garrard's engraving of 'a fat Highland Scotch ox'. The red coat was thick and curly as at the present day but the horns were noticeably shorter. The long horns of modern Highland cattle may have been a development of the later nineteenth century as Youatt (1835) makes no comment on the horns while Low (1845) wrote that the coat should be silky, as it is today, and the horns should be 'of good length, without approaching to coarseness, spreading and tipped with black'. The cattle could be black, brown, mixed black and brown or mouse-dun in colour.

Susie Whitcombe painted the presentation of trophies by HM The Queen at the Royal Highland Show in June 1984 (centenary year of the Highland Cattle Society and bicentenary of the Royal Highland and Agricultural Society of Scotland). The prizewinners are the bull Bhaltair 2nd of Hungerhill, from Sussex, and Mairina 7th of Douneside, from Aberdeenshire.

Shetland

In the early nineteenth century the cattle of Orkney and Shetland were marvelled at for their small size, fat Shetland oxen weighing a mere 4 cwt (200 kg), although Orkney cattle were rather larger. In the opinion of progressive farmers these cattle had no place in agriculture, but they were well adapted to a crucial role in the island economy as small cattle were easier to keep alive in times of hardship and easier to handle in small boats. They existed in remarkably high numbers – Sinclair had reported at the end of the eighteenth century that on the Orkney islands of Sanday and North Ronaldsay, 400 households owned 1570 cattle.

Low, writing in 1845, thought Orkney and Shetland cattle were of Scandinavian origin while Youatt (1835) considered them to be a diminutive race of Highland cattle. Typically cows were milked three times a day to yield about a quart in total. As beef animals they were sold for fattening elsewhere and, until quite recently, cattle were also used for work in Orkney and Shetland although the Shetland breed itself was not used in this way.

The replacement of oxen by horses was a very long drawn out process in Orkney. On the islands of Flotta and Swona, as on Fair Isle, work on the farm passed from oxen direct to tractor in the mid 1940s, without an intervening horse-drawn stage. Indeed, on Fair Isle, before oxen were introduced in the mid nineteenth century, all cultivation was done by hand. From about 1770 to 1830, some improving landlords in Orkney, as elsewhere in Britain, tried to encourage the use of oxen for draught, and they had a certain limited influence but they could not reverse the trend towards horse power.

Hibbert described the Shetland Islands in 1822 and drew attention to the way the four oxen of the typical Shetland draught team were yoked in a staggered arrangement.

One might expect the photographs of the last draught oxen in Orkney, of which there are quite a few, mostly from the 1940s, to reveal links between these cattle and the extinct Orkney breed, but this is not so. The last draught oxen in Orkney were clearly Shorthorn, Aberdeen-Angus or cross-bred. They were not yoked, but wore horse collars, positioned inverted over the neck.

In Shetland unlike Orkney the advantages of the local breed continued to be appreciated, and in 1911 the Shetland Cattle Herd Book Society was founded, publishing Volume 1 of the herd book in 1912, where 380 cows and 39 bulls were registered. Colours were noted and included dun, red, grey, black, and brindle.

In 1958 the Department of Agriculture and Fisheries for Scotland set up the Knocknagael herd near Inverness, to play a part in livestock improvement in the Highlands and Islands. More recently the Rare Breeds Survival Trust has helped with the conservation of the breed, encouraging herds elsewhere in Britain to be set up using stock from Knocknagael and from Shetland (especially from the herds of Mr H. Bowie and Mr T. Fraser), and other private herds have been started. Central to these efforts has been the Shetland Cattle Herd Book Society (revived in 1972, shortly before the Rare Breeds Survival Trust was formed), and financial support given by the Shetland Islands Council.

In September 1983, five in-calf females and a young bull were sent to the Falkland Islands under the aegis of the Rare Breeds Survival Trust, where the breed could have a significant part to play in the reconstruction of agriculture.

'Zetland breed.'
Friesian bulls were crossed with native cattle on Shetland from the 1920s onward but the Breed Society firmly states that pure-bred stocks of Shetland cattle were kept separate and that the black and white markings seen today are due to Shetland breeders having favoured colour patterns like those of the Friesian. Certainly Shiels' painting is indisputable evidence of such markings being characteristic of the breed in the 1840s.

Kerry and Dexter

In the course of his tour of Ireland (1780), Arthur Young commented on the 'poor people's breed of little mountain or Kerry cow' of County Cork and declared that it showed an influence of Alderney cattle. Devon cattle had also had an influence by this time but the virtues of the local breed were not lost. In 1810 the Revd Horatio Townsend, describing the agriculture of Cork, referred to the local cattle as a mixed breed of 'various colours . . . formerly they were all black; in the more remote districts this colour still predominates'. They were very good milkers; 'eight pottles or sixteen quarts a day being no uncommon produce from a cow of three hundredweight'. There was a strong body of opinion in favour of the conservation and improvement of the native cattle, and in 1844 the first classes for Kerry cattle were held at the Royal Dublin Show.

By this time a miniature version of the Kerry, the Dexter, had emerged and one of the first to report on it was Professor David Low in 1845. In his words 'Mr Dexter, agent to Maude Lord Hawwarden, is said to have produced his curious breed by selection from the best of the mountain cattle of the district'. These dwarf cattle have since become more popular in England than in Ireland.

The earliest date for the introduction of the Kerry to England is not known, but in 1882 Dexters were brought over by M.J. Sutton. In the 1886 Royal Show at Norwich a three year old cow was exhibited in the Any Other Breed class. A combined herd book was started in 1887 as a 'Register of pure-bred Kerry cattle and Dexters', taken over in 1890 by the Royal Dublin Society. In the volume for that year there were listed 118 Kerry bulls, 943 Kerry cows, 26 Dexter bulls and 210 Dexter cows. Cattle were also registered by English breeders including Queen Victoria and the Prince of Wales who kept their stock at Shaw Farm, Windsor, and Sandringham, respectively. At first English breeders were content to register in the Irish herd book but in 1892 they formed their own society.

In contrast to the general lack of interest in the Dexter shown by Irish breeders, the Kerry cow received government support in Ireland from the outset. From 1888 to 1902 premiums were paid to encourage the availability of good Kerry bulls for breeding. This was necessary because breed improvement was being hampered by the sale of the best stock to English breeders, and in remote areas farmers continued to use inferior bulls. Under the Livestock Breeding Act of 1925 a Kerry Cattle Area was designated wherein only Kerry bulls could be kept although bulls of other breeds were permitted in some circumstances. These regulations were relaxed after a while but it is striking how government policy had been formulated expressly to help a particular breed.

The Kerry has never been a commercial breed of poorer lands in Britain as it has been in

A six year old cow, the property of the Earl of Clare, painted by Shiels. Close in form to the Kerry of today, Lord Clare's cow (from a stock selected by the Bishop of Killaloe) corresponds to the traditional ideal. However, until the beginning of the twentieth century colours other than black were quite common in the Kerry and Dexter. Volume 1 of the Kerry and Dexter Herd Book includes the entry of a Dexter cow, Lily 2nd, calved in 1887, owned by the Earl of Rosse and coloured (true to her name) 'white with a very little red'. Red Dexters are frequently seen today.

Dexter cattle in the Knotting herd of Miss Jane Paynter at Yelden, Bedfordshire, photographed in 1988. This is one of the few herds of this breed where official milk records are kept, and at thirty head of milking cows it is one of the biggest units. Both black and red colours are represented in this herd which is highly successful at agricultural shows and in supplying stock for export. These cattle have been dehorned for safety in the milking parlour.

south-west Ireland, and the British Kerry is in fact precarious numerically, while the Dexter has been doing well. Both breeds are, for the most part, kept on good land in the south of England.

In 1900 there were 237 registered Dexter females in 15 herds in Britain, rising to 1119 in 69 herds in 1925. The breed contracted to 317 in 24 herds in 1940 but is much stronger now. Pedigree analysis has shown that the Dexter is not notably inbred even though one large and long-lasting herd, the Grinstead herd founded in 1912, had a significant influence on the breed.

Dexters can be black or red while Kerrys are always black. In both breeds some white on the udder or scrotum is permissible. Breeding Dexters is made difficult by a tendency for 'bulldog' (achondroplastic) calves to be produced. These genetically deformed calves are always born dead and are more common from 'dwarf' parents (for a general description see Alderson, 1978).

The Kerry has changed very little over the last century. This fine bull with metal finials on his horns dates from 1856 and was figured by Baudement (1861).

36

Irish Moiled

'The Polled Irish Breed, scarcely known to the breeders of England' was, according to Low (1845) light brown in colour and found throughout Ireland, being most abundant around the river Shannon. It was said to resemble the Suffolk Dun. By this time Shorthorn cattle had established their ascendancy over most of Ireland and at the start of the twentieth century the Irish Moiled was associated only with the three northern counties of Tyrone, Armagh, and Sligo. In 1926 the Irish Moiled Cattle Society was formed in order to develop the breed along dual-purpose lines for the small hill farms of Ulster. Capt. Herbert Dixon and Capt. J. Gregg were President and Secretary respectively.

The new breed attracted considerable support, with classes being held at the Royal Ulster Show, and a bull premium scheme was started whereby good Irish Moiled bulls were made available for small farmers. In 1929 a preferred colour standard was laid down; red or roan with a white stripe down the back, a white tail and white underparts. However, when Capt. Gregg died the Society fell into a decline which was only reversed when it was reconstituted in

Photographed by Mr E.J. Boston in the 1950s, this Irish Moiled bull, Maymore IV, is wearing a metal face shield to prevent him from attacking people.

1948. The next year Maj. G. Perceval-Maxwell formed his Ballydugan herd and in 1950 a polled bull called Hakku was imported from Finland. Capt. Gregg had believed firmly that the Vikings stole Irish Moiled cattle for breeding, giving rise to the polled stock of Scandinavia, and that this was therefore an appropriate source of new blood. Hakku had a wide influence on the breed.

New Government regulations meant that only bulls whose dams had a high recorded milk yield could be licensed for breeding and most breeders of Irish Moiled cattle did not keep such records. Again the breed went into decline, and in the early 1970s there were only about thirteen pure-bred females and six bulls, with two breeders, Mr David Swann of Dunsilly and Mr Nelson of Maymore, keeping the breed going. A revival took place once more with the reconstitution of the Breed Society, and in 1984 Irish Moiled cattle were again exhibited at the Royal Ulster Show, forty-five years after the previous classes had been held.

On 6 March 1984 the British Post Office released a special set of five stamps featuring native cattle. This 31p stamp, featuring the Irish Moiled cow, has in the background the Legananny Dolmen, a megalithic tomb near Ballynahinch, County Down, believed to date from 3000 BC.

Welsh Black

> Welsh cattle are extremely various; every province in the Principality seems to send out a separate breed . . . they vary, in regular gradation, from the largest ox to the lowest runt.

In the 200 years since Marshall wrote this (1789), the Welsh Black has become firmly established as a breed. This has not been the work of a single great improver, or group of breeders, but the result of experimental introduction of blood from other breeds against a background of a husbandry system based on the tenanting of small farms. Where the land was good enough for turnips to be cultivated for winter feed, notably in Brecon, Radnor, and Montgomery, the Hereford and Shorthorn breeds took over completely. Improvement of the local stock was slow, principally because good beasts tended to be sold for the droving trade and were not used for breeding.

Early in the nineteenth century, Welsh cattle in the south and east of the country were extensively crossed, initially with Improved Longhorns, latterly with Herefords, Devons, Gloucesters, Shorthorns, and Dutch cattle. Three main races emerged; the Pembroke (later known as the Castlemartin), the Glamorgan, and the Carmarthen. In Merioneth a polled variety existed and in the 1950s about forty females of this strain were used as a nucleus for the Poll Welsh Black. The 'Anglesey runts' which had been the mainstay of the droving trade remained relatively pure.

The Pembroke (painted by Shiels) had only emerged as a distinct type in the eighteenth century; it acquired a good reputation for dairying and beef in southern England as well as on its home ground. Morgan Evans, writing about the breed in 1881, said it was cheaper to feed and rear than the Hereford or Shorthorn which needed to be housed for three weeks longer in winter. The Castlemartin and the Dewsland were varieties of the Pembroke.

Welsh Black cattle are prominent in this landscape, painted near Aberystwyth by Walter Bayes (1869–1915).

Later in the century, the Welsh types began to merge and in 1874 the Welsh Black Cattle Society was formed. The Earl Cawdor, who was prominent in the Society, had misgivings about including the North Wales types as he thought they would be better served by having their own herd book and they did so from 1883 to 1905.

Colour and conformation were not standardized at all in the early years of registration. The 1883 volume of the herd book reveals that 6 females were black and white, 3 were all red, and 2 were red and white. Milking ability, carcass quality, and hardiness were of equal importance for the breeders, and today the Welsh Black is remarkable for combining these features to the extent that until recently it was regarded as dual-purpose. Neither of the other important hill breeds (the Highland and the Galloway) can be said to have this quality.

Although Welsh Black herds were set up in England and Scotland, the breed remained local to Wales and it was only with the grant-aid to hill farming which began, effectively, in the 1950s that hill cow numbers began to increase markedly. In the twenty years to 1976, the number of cows for which subsidy was paid rose from 44 000 to 148 000. The market, too, began to favour the breed. There are now about 70 000 breeding Welsh Black cows in Britain but, unusually for such a numerous breed, there are very few abroad.

There is a number of colour variants of Welsh Black, including white and belted patterns, although the breeding of these has not been officially encouraged.

Daniel Clowes painted these Welsh cattle, Benthal and Bran, in 1825, for their owner, Sir Robert Vaughan of Nannau. The North Wales type was of adequate beef conformation, and was the mainstay of the Welsh droving trade. Sir Robert favoured the development of milking qualities, and is known to have imported Dutch cattle for this purpose. In this picture the herdsman carries yoked milk churns; the brass horn finials should be noted.

White Park

Chartley bull, by Charles Tunnicliffe RA (1901–1979). Chartley cattle lived in Chartley Park, home of the family of the Earl Ferrers, till 1904, when tuberculosis reduced numbers to eight or nine. The herd was sold to the Duke of Bedford at Woburn and crossed with Longhorns, with some subsequently going to Whipsnade. The Woburn cattle were sold to Lord Ferrers, now resident in Suffolk, in 1971, and were soon joined by the group from Whipsnade.

Fallow deer were the main big game animal for the barons and kings of the middle ages, but for a handful of magnates there was a yet more thrilling quarry – the wild bull. The wild cattle of Windsor Forest were referred to in 1277; those of the Stirlingshire area were mentioned by Hector Boece in the sixteenth century, and there are several other records. However, the link between the Medieval herds and the confined park herds (which were first documented in the seventeenth century) is not clear. While it is possible that local herds of free living cattle were herded into the parks of Chillingham, Chartley and Cadzow when they were enclosed, the evidence for this is purely circumstantial.

The Chillingham herd is the only one still living in the conditions traditionally associated with such cattle. So romantic and well known had they become in the nineteenth century that several other herds of similar cattle were started, or their existence publicized. The Cadzow herd, which according to tradition had been inhabitants of the Duke of Hamilton's park in central Scotland since time immemorial, acquired particular fame through the romantic lyricism of Sir Walter Scott:

> Mightiest of all the beasts of chase
> That roam in woody Caledon,
> Crashing the forest in his race,
> The mountain bull comes thundering on.

Again with the exception of Chillingham, all the herds of white park cattle exchanged stock and used cattle of other breeds directly or indirectly at some time or other. In 1918 the Park Cattle Society was founded, with the Earl of Tankerville (Chillingham) as first Patron, the Duke of Bedford (Woburn) as President and the Duke of Hamilton (Cadzow) as Vice-President. Nine horned and six polled herds were listed in Volume 1 of the herd book. The British White split off and adopted its name in 1946 while the White Park Cattle Society continues to represent the horned breed, although the Chillingham herd has not been considered part of it since 1932. It is misleading to refer to Chillingham cattle as White Parks; they have the status of a breed in their own right.

The Vaynol herd was started in 1872 when cattle descended from the wild herd at Blair Atholl, Perthshire, (crossed with white Highland cattle) were moved to Vaynol Park in North Wales. On the death of their owner, Sir Michael Duff, in 1980 they were transferred to Shugborough Park Farm in Staffordshire and thence in 1983 to the Rare Breeds Survival Trust. Then they comprised seven cows and their calves, a mature bull and two young bulls. The Dynevor herd, from near Llandeilo in south-west Wales has no recorded history prior to 1898, but although it, and the Vaynol herd, may have only short written histories circumstantial evidence favours the idea of long association of each of these herds with baronial

White cattle in front of Streatlam Castle, County Durham, the painting attributed to Joseph Miller and probably painted in the first half of the nineteenth century. Could these cattle be the subject of artistic licence? There is no record of a white herd at Streatlam, although there were herds of white cattle nearby in County Durham at Bishop Auckland Park and Barnard Castle.

A White Park bull's head forms

A White Park bull's head forms the emblem of the Rare Breeds Survival Trust and bulls of this breed are popular exhibits at rare breed events. Here at the Autumn Exhibition 1986, East of England Showground, is Chartley Michael. His pedigree shows how present day White Park cattle are the result of crossing among several ancestral herds, and that he is inbred. Dynevor Samson was his sire's grandsire as well as his dam's sire.

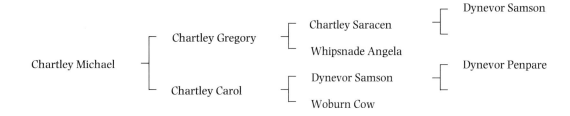

hunting parks and castles, many of which were created in the thirteenth century (Whitehead, 1953).

On many occasions in the last 200 years the different herds of White Park cattle have given cause for concern, either through outbreaks of disease or through loss of viability attributed to inbreeding. To bring in fresh blood, bulls from Chillingham, Dynevor, and Vaynol have been used in the Cadzow herd. The Chartley herd was rescued from extinction by being crossed with Longhorns and a Dynevor bull was used. The Dynevor and Vaynol herds have been crossed with British White and Cadzow bulls respectively. Only the Chillingham cattle remain pure, and they are discussed in the next section.

White Park cattle today often have coloured calves, which until recently were always culled at birth. As yet it is difficult to see a commercial application for White Park cattle, but the breed society is encouraging the use of the bulls as crossing sires. The main value of the breed continues to be decorative, as it always has been.

'Wild or White Forest breed. Cow, eight years old, from Haverfordwest in the County of Pembroke'. The Dynevor cattle are probably associated with white cattle of this kind, which were themselves of the stock which gave rise to the modern Welsh Black. This painting by Shiels was reproduced as a plate in Low's book. In Low's words: 'The individuals of this race yet existing in Wales are found chiefly in the county of Pembroke, where they have been kept by some individuals perfectly pure, as a part of their regular farm stock. Until a period comparatively recent, they were very numerous; and persons are yet living who remember when they were driven in droves to the pastures of the Severn, and the neighbouring markets. Their whole essential characters are the same as those at Chillingham and Chartley Park, and elsewhere . . . Individuals of this race are sometimes born entirely black, and then they are not to be distinguished from the other cattle of the mountains.'

The Chillingham herd offers a unique opportunity to study the behaviour of a herd of cattle with a natural sex ratio and age structure. These cattle breed all year round. The cows calve away from the herd but then return to it, and only visit the calf to suckle it every six hours or so for the first ten days. After this time the calf may follow its mother back to the herd. If female, it remains in the herd for life. If male, when it reaches the age of four years it leaves the herd and joins, or establishes, a home range with one or two other bulls of similar age. These home ranges persist for life. Numbers in the 1980s have been around the middle fifties.

Chillingham cattle are white, although many have distinctive patterns of black or red-brown spots on the neck, face and shoulders which enable individuals to be identified. Horn shape is individually distinct in many females, but not in males. Bulls probably weigh 300 kg, cows 280 kg, with shoulder height no more than 110 cm. This represents a small animal compared to other breeds. Sexual maturity comes late and heifers do not usually bear their first calf before the age of four years.

Chillingham

Though dedicated to the cause of progress many of the eighteenth century agriculturalists retained an affection for the glories of bygone days. John Bailey, who with George Culley compiled the *General View of the Agriculture of the County of Northumberland* (1794) was steward for the Earl of Tankerville at Chillingham, Northumberland, and he took a great interest in the white, horned, red-eared cattle which had lived in Chillingham Park since time immemorial. Bailey's account of their natural way of life, and how in former times they were hunted from horseback, was printed in Thomas Bewick's book of 1790.

Charles Darwin prompted the keeping of records of the herd and from 1862 to 1899, births, deaths, numbers, and 'remarkable occurrences' were dutifully noted. From these records it emerges that 40 per cent of bull calves were castrated. A number of cows were shot, mostly to terminate suffering, but 12 per cent of the cows shot were killed for their beef. Presumably it was the barren cows that were shot for beef and the poorer bull calves that were castrated indicating that the herd was under strong artificial selection at this time. Winter feeding of the herd was mentioned in a letter of 1721 and probably took place for a long time before that. Today such hay feeding is the only overt human interference with the herd; culling was last practised in 1918 and no bulls have been castrated this century.

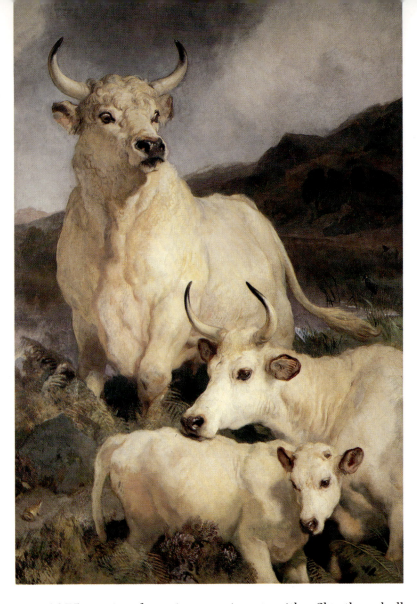

Sir Edwin Landseer (1802–1873), painter of this picture, and Sir Walter Scott (1771–1832) were among many who helped to romanticize the Chillingham cattle until the story gained credence that they are in some way a particularly direct descendant of the aurochs, the wild ancestor of domestic cattle which died out in Britain in pre-Roman times. It is far more likely that the Chillingham herd represents a relic Medieval stock enclosed in Chillingham Park some time between its establishment in the thirteenth century and 1645, the date of the earliest written reference to wild cattle at Chillingham. The founder stock might have been roaming Northumberland or the Scottish Borders in a feral state, although this is unlikely, or may have been specially selected from domestic herds. This picture was first exhibited in 1867.

In 1875 a series of crossing experiments with a Shorthorn bull was tried, and these proved more successful than the attempts to cross Shorthorn heifers with one of the wild bulls. Two families were created; Chillingham Wild Rose and Chillingham Wild Blossom, and progeny of these crosses (none of which was returned to the wild herd) were successful at Smithfield. One bull, Albion, was eligible, as the result of four generations of crossing, for registration in *Coates Herd Book*.

The Tankerville estates were badly hit by the depression at the end of the nineteenth century, but the preservation of the wild cattle had become a family tradition and in 1939 the 8th Earl, with help from a committee of prominent Northumbrians, formed the Chillingham Wild Cattle Association. The cattle, and depasturing rights over Chillingham Park, were leased to the Association. On Lord Tankerville's death in 1973, ownership of the cattle passed to the Association. The park itself was acquired by the Sir James Knott Charitable Trust in 1981 and the lease was extended to 999 years. Additionally in 1973 a reserve herd was established in the north of Scotland.

British White

These cattle are descended from the Middleton Park herd (near Bury, Lancashire) which is traditionally supposed to have been founded with cattle from Whalley Abbey nearby. Around 1765 the herd was moved to Gunton Park in Norfolk, and it has had a strong association with that county ever since.

Another link between the early history of the breed and that of the present day is the Somerford Park herd in Cheshire which, although there are no written records, is presumed to have originated from Middleton Park cattle sometime in the seventeenth century. Before it was dispersed in 1925 Shorthorn and Chartley bulls had been used. Vaynol and Cadzow blood has also entered the breed, via the Faygate herd (Sussex) which was founded in 1908 and which is still active.

The history of the breed has been chronicled by Whitehead (1953). As in the White Park, several long-established herds have kept the breed going even while they cannot have paid their way. Even in 1951, not a time when rare breeds were fashionable, there were 11 herds of British Whites with a total population of 642 head; by 1985 there were 36 herds with about 780 registered cattle. This was the year that classes for the breed were reinstated at the Royal Norfolk Show after an absence of twenty years. Today export markets, notably to Australia and the USA, are strong. Most if not all British Whites are kept as beef cows, while much effort is being put into promoting the breed as crossing sires for other breeds. The British White used to be dual-purpose but the last dairy herd, the Hevingham herd in Norfolk, gave up milking in 1974.

W.A. Clark painted this British White bull, from the Woodbastwick herd. The bull was first in his class at the Royal Show, Cambridge, in 1922.

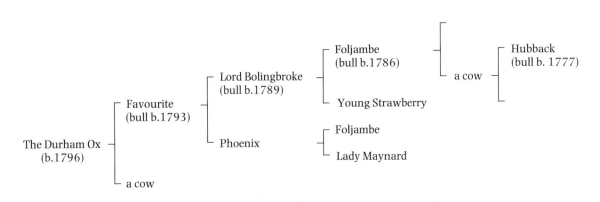

3 Short-horned cattle

It is probable that short-horned cattle were imported from the Netherlands in some numbers at the end of the seventeenth century. They were first given the name of Holderness after the district in Yorkshire where a variety became established. Later, towards the end of the eighteenth century they became known as Teeswater cattle, after the district where the famous Colling brothers lived. Charles and Robert Colling were responsible for such skilful improvement in the breed that the Shorthorns overtook the Longhorn as the favoured breed for both beef and milk, and so they remained throughout the nineteenth century.

Like Bakewell, the Colling brothers achieved their improvement by intensive inbreeding. They began in about 1780 when they obtained the great sire Hubback who was descended from the Studley bull (calved 1737); himself a medium sized descendant of a Low Countries strain. This line of Teeswater cattle culminated in the Durham Ox, a remarkable animal that was probably the most famous of all the Shorthorns. It was bred from a common cow put to the famous bull Favourite (see pedigree). At five years old, in 1801, the Durham or Wonderful Ox was sold by the Collings brothers for £140 to Mr Bulmer who travelled with it, on display in a carriage, for five weeks. At that time the ox weighed 3204 lb (1456 kg). Mr Bulmer then sold the ox and its carriage, on 14 May 1801, to Mr John Day for £250. Youatt (1835) wrote that on the same day the ox could have been sold for 500 guineas, on 13 June for £1000 and on 8 July for £2000. Mr Day declined these offers and travelled with the ox, exhibiting it to the public for nearly six years until it was finally slaughtered in 1807, after dislocating a hip.

Unimproved short-horned cattle were of great size and heavy yielders of milk but did not breed well. The improved breed was much neater, although still large, and with good beef qualities. Later, Thomas Bates (1775–1849) attempted to redress the loss of milk that was a result of the improvement by the Collings and their followers. Shorthorns took a long time to recover their eighteenth century reputation for milk yield; it was not until the latter part of the nineteenth century and with scientific feeding that the Dairy Shorthorn gained a rather short-lived supremacy before being overtaken by the Friesian.

This is the Durham Ox, also known as the Wonderful Ox. Garrard claimed that it was 'no less distinguished for its uncommon beauty than for its weight, being an example of perfection in every way'. In 1802 when Garrard made this engraving the ox stood 5 ft 5 ins (165 cm) at the shoulder and had a girth over the ribs of 10 ft 10 ins (330 cm).

While the Colling brothers were establishing the improved and highly bred Shorthorn in the north of England, the Scottish Shorthorn, today known as the Beef or Scotch Shorthorn, was being developed as an important beef breed. The breeders were not so interested in pedigree as in producing cattle of the right conformation, with a low chest, deep ribs, and a wide back. During the middle of the nineteenth century Scottish Shorthorns were used to cross with the English breed to improve the inbred descendants of the Colling's stock.

Shorthorn

More has been written about the Shorthorn than about any other breed of farm livestock. For a hundred years, up to the 1950s, it was the most numerous breed of cattle in Britain, and today the Shorthorn is the most widely distributed breed in the world. Most of the highest prices for pedigree livestock have been paid for Shorthorns, ever since the Collings' bull Comet was the first bull to be sold for a thousand guineas, in 1810. The Shorthorn has the oldest herd book (apart from the *General Stud Book* for horses, founded in 1791), known as *Coates Herd Book*, which was started in 1822, and the oldest breed society, the Shorthorn Society of the United Kingdom of Great Britain and Ireland (founded in 1872).

In the first detailed cattle census of 1908, $4\frac{1}{2}$ million of Britain's 7 million cattle were described as Shorthorns. How the Shorthorn acquired its pre-eminence is a fascinating piece of economic and social history, and the recent studies of J.R.Walton (1984) show how this can be investigated. Certainly the history of pedigree Shorthorns has been very extensively studied, in terms of 'tribes', 'lines', championships, trophies and record prices, but never with any objective analysis of the effects of the money and effort expended upon increase of yield of beef or milk, let alone of the efficiency of such a process.

It is with the Shorthorns that the extraordinary world of the nineteenth century livestock breeders is best revealed. The stories about prize bulls and their breeders are legion, beginning with that of Comet, the most famous of all bulls. Comet was bred by Charles Colling and was sold, as a six year old, to a syndicate of four breeders at the Ketton sale where the herd was dispersed on Colling's retirement in 1810. He was a light roan, and was sired by Favourite out of Young Phoenix, who was herself the daughter of Favourite having been mated to his own mother. As H.H. Dixon (1868) wrote, 'He was not very large, but with that infallible sign of condition, a good wide scorp or frontlet, a fine placid eye, a well-filled twist, and an undeniable back. His price caused breeders everywhere to prick up their ears'.

Comet was the sire of Duchess I, purchased by Thomas Bates at the same sale. The Duchess line became particularly famous; Duke of Northumberland, a bull out of this line, won first prize at the Royal Show in 1839. Bates-bred Duchess cattle were very important in the development of the North American Shorthorn and while not prolific they acquired esteem for their milking qualities. By 1873 the Duchess line had died out in Britain; all available cattle of this breeding were in the USA and were put up for auction that year near Utica, New York. Fabulous prices were paid with one cow selling for 40 600 dollars (then £8337).

Shorthorns have always had a reputation for being very prepotent, that is they are highly effective in bestowing their characteristics on their progeny. In North and South America in particular, this prepotency was exploited in the upgrading of range stocks of hardy, long-

One of the first engravings of cattle to be published; made by Cuit (1780). It depicts an ox, bred and fed by Christopher Hill of Blackwell, Co. Durham, and slaughtered as a six year old in 1779. The carcass, comprising 151 stone 10lb of beef on the bone and 11 stone of tallow [with 14lb or 6.35 kg to the stone] was sold for £109 11s. 6d. The live weight of the beast must have been at least a ton and a half (1500 kg).

Holderness (Shorthorn) cattle were not favoured by Garrard. They were the largest cattle in Britain at the end of the eighteenth century. This bull had a shoulder height of 5 ft (152 cm).

In Garrard's time the Holderness was still the most common provider of milk, giving 'eight or nine quarts at a meal [sic]'. This cow had a shoulder height of 4 ft 11 ins (150 cm).

Another print of the Durham Ox shown on p. 48. About six times as many different engravings were published of Shorthorns as of any other breed. This was the most popular, engraved by Whessell after Boultbee, of which more than 2000 copies were sold within a year of publication in 1802. After 1845 fewer livestock pictures were published; why they should have gone out of fashion is not clear.

Garrard claimed that many of the oxen 'were large in the bone, heavy and inactive, consequently slow feeders, and by no means so well calculated to purposes of husbandry as the middle-horned breeds'. This Holderness ox had a shoulder height of 5 ft 6 ins (168 cm).

Winbrook Atom 2nd, a Northern Dairy Shorthorn bull, at the Great Yorkshire Show in 1974. This type of Shorthorn was developed in the Pennine dales and had its own breed society from 1944 to 1969. Shorthorns of this hardy dual-purpose type continue to be registered in rather small numbers in *Coates Herd Book*. Four bulls of this breed, including the one illustrated, are represented in the Rare Breeds Survival Trust semen bank, which includes straws of semen from 26 other bulls of, in total, seven breeds. The establishment of this bank, which represents the ultimate reservoir of genetic variability for these breeds, was one of the first actions of the Trust.

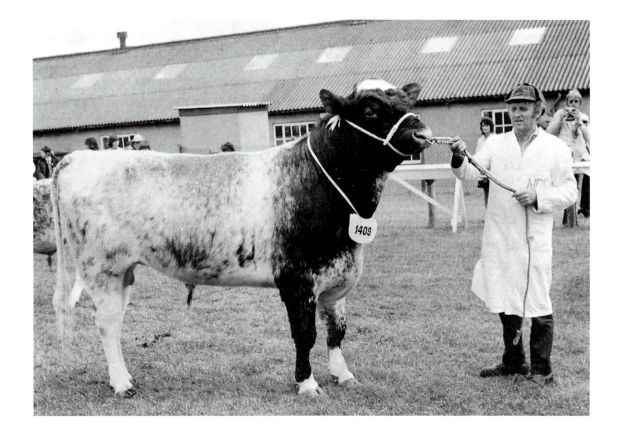

horned cattle of Spanish type, the descendants of those brought by Columbus's followers. These cattle were developed into fine beef animals and their carcasses were sold on the world market.

By 1900, cheap imported beef of high quality was freely available in Britain and the emphasis of British cattle raising had to change to dairying. In times of hardship it is more efficient to use cattle to produce milk rather than beef, so that the economic pressure for change was greatly increased during and between World War I and World War II. This acted against the dual-purpose breeds and favoured the Friesian, that soon took over as the most popular dairy breed. The Friesian also became the source of most of the nation's home-grown beef, through the use of beef bulls on the cows. The Shorthorn was not popular for crossing with dairy cows because it did not colour-mark, and because one of the advantages of the breed as a crossing sire, namely its conferring of milking abilities on its offspring, was not advantageous when the dam was herself a milk cow.

Most present-day breeders of Shorthorns take great pleasure in the fact that they are perpetuating a breed which is more thoroughly rooted than any other in the cultural history of modern British farming. For such enthusiasts, pure-breeding is all important, but it is worth remembering that the Shorthorn does have certain very well-established specialist roles as a crossing sire on beef cattle. The 'Irish stores', so significant a part of the British livestock scene this century, have always had a great deal of Shorthorn ancestry, and it was the importance of this trade that led the Irish government, in the 1920s and 1930s, to discourage the spread of the Friesian in order to protect the Irish herds of Shorthorns. The role of the present-day Shorthorn in cross-breeding is most clearly understood in the Blue-Grey, where the Whitebred Shorthorn is crossed with the Galloway cow to give a milky and hardy suckler cow which is today generally mated with a Charolais or similar bull to give the slaughter generation (see pp. 28–29).

Breeding Blue-Greys for higher milk yield, while not losing hardiness, would be well worth while. However, Whitebred Shorthorns (the sire of the Blue-Grey) tend to be bred by one set of breeders, Galloways by another, and the Blue-Grey heifers tend to be sold to yet another set of farmers. As a result it is difficult to identify a Whitebred Shorthorn bull whose daughters perform particularly well and selection of bulls has tended to be on type rather than on performance.

The crossing of Highland cows with Beef Shorthorn bulls greatly improves their efficacy as suckler cows. The Luing breed was developed in order to combine the respective merits of the two breeds, in a type which could be bred pure. The original aim was to enable hill-cattle farmers to establish permanent herds, rather than to depend on buying in replacements (even with a herd life of ten years, 10 per cent of the stock will need to be replaced every year). The Cadzow brothers of the island of Luing in the Inner Hebrides began developing this new breed in 1947 and it has achieved wide recognition in Britain and abroad.

Shorthorn breeders have tried to modernize aspects of the breed by introduction of blood

James Bateman (1893–1959) painted a Shorthorn bull being delivered to market against a backdrop of Lewes Castle, Sussex. At the time (1937) the Shorthorn was still the most important breed of cattle in Britain. In 1937–8, 23 730 Shorthorn bulls were licensed in England and Wales against 2671 Friesians. Thirty years later the application of AI had led to a dramatic reduction in numbers of breeding bulls and there were 4614 licensed Friesians and 365 Shorthorns. Coincident with dairying becoming the dominant sector of the livestock industry the specialist dairy breed replaced the dual-purpose.
At the other end of the country, the Cumberland and Westmorland Shorthorn retained a place in cattle husbandry. It had become distinguished before the mid nineteenth century on account of its dual-purpose character. Dairying qualities were favoured, in contrast to the feeding for beef prevalent elsewhere.
The Whitebred Shorthorn had its origins in Cumberland. In 1962 the Whitebred Shorthorn Association was formed to provide registration facilities for local breeders who for a hundred years had been developing a race of Shorthorn with milking qualities and little or no red coloration for the specialized purpose of siring Blue-Greys out of Galloways.

from other breeds. In 1970 an experiment on the introduction of genes from the Red Friesian, Red Holstein, Danish Red, Meuse Rhein Ijssel, and Simmental was started and now many Dairy Shorthorns are officially designated Blended Red and White Shorthorns. In 1972 a crossing programme with the Maine Anjou was begun by a private breeder of Beef Shorthorns aimed at adding carcass size to the breed.

The family Stirling of Keir was among the leading Scottish breeders of Shorthorns in the mid nineteenth century. This picture by James Macleod is unusual in that a beef cow is depicted with her calf. In many beef herds calves were put on to nurse cows to preserve their natural dams in good condition for show; according to H.H. Dixon (1868), allusion to milking qualities was often considered a veiled apology for poor beefing characteristics.

The most famous Scotch Shorthorn breeder was Amos Cruickshank who took the tenancy of Sittyton, near Aberdeen, in 1837. He developed the beefing qualities of the breed to a high degree. The private dispersal sale of his herd in 1889 and the success of the bull Field Marshal in the Royal herd at Windsor helped the Scotch Shorthorn to establish itself with English breeders.

Today the pure-bred Beef Shorthorn is numerically rare, with only about 1000 females. This is partly a reflection of its function as a crossing sire, with a perfectly healthy demand for bulls being met from a small breeding population of cows. Likewise the Whitebred Shorthorn is scarce. The Dairy Shorthorn is in absolute terms numerically strong with about 10 000 cows, but tiny when compared with the Friesian at 2.5 million.

Blue Albion

Crosses of black cattle with white cattle often result in a blue roan coloration. Such blue cattle were favoured in the Midland counties of Derbyshire and Staffordshire at the beginning of the twentieth century and were often entered in the local shows. They could have arisen from crosses of Welsh Black and Shorthorn, or from crosses of Friesian and Shorthorn.

In 1920 the breeders of blue cattle felt confident enough to found the Blue Albion Cattle Society and they purchased the Herd Book from William Clark who had begun registering pedigrees of these cattle in 1916. In 1921, when the Royal Show was held at Derby, classes were provided for Blue Albions. The new breed attracted a very great deal of interest, and Volume 3 of the Herd Book listed 117 bulls and 162 females, the progeny of registered parents, as well as 3237 foundation cows registered after inspection had certified each as a 'really good example of a Blue Shorthorn'.

It seems that many, if not most, breeders were unaware that the blue colour could not be made to breed true. Had the genetics of cattle coat colour been understood by the Blue Albion breeders they would have realized that on average only 50 per cent of blue bull × blue cow matings would give a blue calf. Having made blue coloration a breed stipulation the Society had sown the seeds of its own destruction.

The foot-and-mouth disease outbreaks of 1923–4, the worsening agricultural depression during the 1920s, and at least one scandal involving the fraudulent replacement of an incorrectly marked pedigree calf by a blue calf bought in the local market, also contributed to the demise of the Society, which held its last Annual General Meeting in 1940. It was formally wound up in 1966.

Today two herds exist which may be connected with cattle registered as Blue Albions before 1939. There is no breed society and the priority attached to the conservation of these cattle has been low, although the Rare Breeds Survival Trust does store semen from three bulls.

The Ryleys herd at Alderley Edge, Cheshire, has been claimed to have links with pre-war registered Blue Albion cattle. They are a striking sight in their parkland setting.

Lincoln Red

The 'high farming' of Victorian times, in the east of England, East Lothian, and north-east Scotland, was based on the firm integration of crop and livestock husbandry. Cattle were fed in stalls, and their rich manure carted to the fields. Philip Pusey, brother of the famous churchman and a leading commentator on agriculture, wrote in 1842 'the stock which furnished our forefathers with meat were fattened on rich grasslands but . . . by a great revolution in farming the light arable soils now chiefly supply the country with animal food.' In Lincolnshire, dependence on home-grown fodders for cattle fattening was rather low, farmers preferring to feed sheep off the turnip crop in the field because of the lightness of the soil, and oilcake was bought in for the cattle. The Shorthorn responded well to this kind of feeding and the variety which found favour was that which had, from the late eighteenth century, been selected by Thomas Turnell of Wragby to combine beef and dairy qualities, and an even all-red colour. The first volume of *Coates Herd Book* distinguished this variety, in 1822.

In 1895 the Lincoln Red Shorthorn Society was founded and the breed became well known for its dual-purpose qualities, but was practically never seen outside its native area. This was

Lincoln Red Cattle Society Field Day at Manor Farm, Anwick, Lincolnshire, on 14 August 1963. The polling factor had clearly not taken over in this herd, founded in 1908. By 1976 only a few horned Lincoln Reds were to be found, but since then, the continental additions have resulted in horns reappearing.

No-one knows how Turnell bred his cherry-red cattle, nor how much their quality owed to the indigenous nondescript cattle he must have crossed with his bought-in pedigree Shorthorns. Certainly Defoe, writing about Lincolnshire in 1724, reported that almost all the cows for a distance of fifty miles were red, or red and white. This heifer, one of twins, measured 8 ft (244 cm) from horns to tail; bred by Mr Harrison of East Keal, and fed by Mr Benniworth of Toynton, she was painted by J. Tennant around 1810.

in sharp contrast to the Beef (Scotch) Shorthorn which became famous world-wide. However, in at least three respects the Lincoln Red has been a pioneer. In 1956 the breed society began promoting the polled variety, which Eric Pentecost had begun to develop in 1939, bringing in the genetically dominant polling factor from the Aberdeen-Angus. Perhaps to emphasize the new identity the Society dropped 'Shorthorn' from its name in 1960. The next year, in advance of all other British breeds, it set up a beef recording scheme, animals to be weighed at weaning and at about 400 days with a target weight of 550 kg. It was the first society to publish official weight figures in bull sale catalogues and to stage special classes for weight recorded animals. In 1977 the Breed Development Scheme was started to increase size by bringing in blood from the Maine Anjou and other continental breeds. Today there are about thirty-two registered herds in Britain.

The Lincoln Red has become established in Australia, South Africa, Brazil, and Argentina, but Canada is where it has done best. Strangely enough, this was a result of the 1959 revolution in Cuba; a consignment of the breed for that country was trapped in Canada by the US trade embargo on Castro's government and was sold locally.

4   English lowland cattle

In the late eighteenth century cattle were classified according to the length of their horns, as long-horned, short-horned, middle-horned, or polled, as well as three groups that were described separately; the Channel Island breeds, the Scottish breeds, and the wild stock (see previous chapters for the two latter groups).

Amongst the lowland cattle of England there were breeds of all three horn lengths as well as polled breeds, and these were often interbred, especially with the improved Longhorn. The Longhorn was an ancient and distinctive race of domestic cattle going back to the Medieval period. It was a long-legged animal with a thick hide, long, straight horns and of various colours but most individuals had the white stripe running down the spine, the 'finch-back', that is still characteristic of the breed today. It was this breed that was improved by Robert Bakewell, and by the end of the eighteenth century it had become as widespread throughout England as the Friesian has been in this century.

The Shorthorns have already been described (pp. 49–59). The middle-horned cattle comprise the main group of the indigenous English breeds; the Devon, South Devon, Hereford, and Sussex. Pre-eminent amongst these in the eighteenth century was the Devon which had the reputation of being the finest draught oxen in the country. This breed represented the old generalized type of red cattle from which all the red middle-horns had been developed.

Amongst the Polled breeds in the Lowlands were the Red Polls of Norfolk and Suffolk which were probably of ancient origin, although these cattle could well have been interbred with polled cattle from south-west Scotland, driven in their thousands through East Anglia to the meat markets of London. The indigenous East Anglian polled breed was the Suffolk Dun which was a dairy cow.

From a painting by Susan Crawford of Androsspoll, a Hereford bull.

Longhorn

By the mid 1800s, Longhorns were to be found only in a few herds, such as that of Sir John Harpur Crewe, Bt., at Calke Abbey in Derbyshire. This picture (left) was painted in 1880 by W.R. Woods. The Abbot of Calke and Canley 2nd are in the foreground; Lofty 2nd is the middle one of the three cows further off, with Tulip 10th on the left and Beauty 4th on the right.

Although this engraving of a Longhorn bull is dated 1800, Garrard wrote in his text that the print 'was executed from a picture made in January 1796 from Mr Honeyborn's bull called G.N. then three years old.' Measurements of this bull were not recorded.

Cattle of all horn types in the English midlands of the early eighteenth century were used for draught as well as for meat and milk. Locally produced beef came from old draught oxen, but with the increase in the city populations a need developed for breeds specifically intended for beef production.

By the middle of the eighteenth century several breeders were beginning the systematic improvement of local stock. The most famous, of course, is Robert Bakewell, of Dishley, Leicestershire, who by 1760 was gaining a reputation as a sheep breeder. He bought two Longhorn heifers from another notable improver, Mr Webster of Canley, near Coventry, and a bull from Westmorland, whose offspring he inbred intensively to produce early maturing animals with a carcass that had a high proportion of fat and relatively little bone; he was not concerned unduly with milk yield, and one of his sayings was 'all is useless that is not beef'. His cattle were expected to fatten to between 560 and 840 lb (254 to 381 kg); their horns measured from 15 inches to 2 feet (38 to 61 cm) for bulls, and from 2 feet 6 inches (76 cm) for oxen, while the cows' horns were nearly as long as those of oxen but finer and more tapering.

The fame of Bakewell's Dishley Longhorns, also known as the Improved Longhorn or New Leicester breed, spread rapidly. This is the earliest example in cattle of a breed structure where a few famous, long-established herds provided bulls highly sought after for the use of the breed as a whole and for the upgrading of local nondescript stocks. As a result, by 1810 the overwhelming majority of cattle in the midland counties (particularly Leicestershire, Derbyshire and Staffordshire) were of Improved Longhorn type. Longhorns also contributed to the Australian cattle herds, and a few went to the USA in 1817 and subsequent years, but they had very little influence there. The Texas Longhorn was quite unrelated to the British in its origin, being descended from Spanish stock.

'Improved long-horned or New Leicester cow.'
This cow was of Dichley stock (as spelt by Garrard). It had a shoulder height of 4 ft 1 in (125 cm).

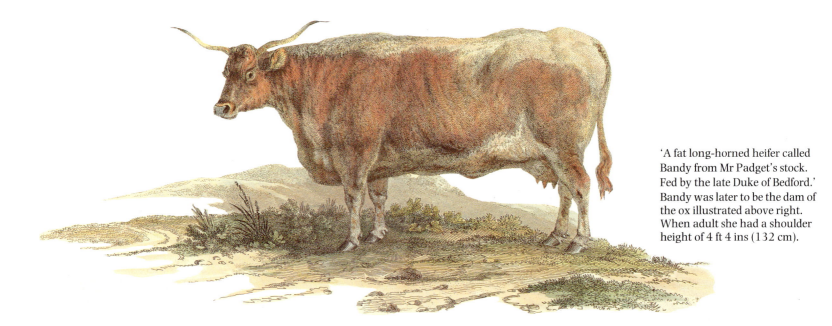

'A fat long-horned heifer called Bandy from Mr Padget's stock. Fed by the late Duke of Bedford.' Bandy was later to be the dam of the ox illustrated above right. When adult she had a shoulder height of 4 ft 4 ins (132 cm).

Garrard wrote that this 'improved long-horned or New Leicester ox' was 'in the possession of the Duke of Bedford, bred by Mr Walton of Ibstock, off Bandy, by Mr Padgets Shakespeare'.

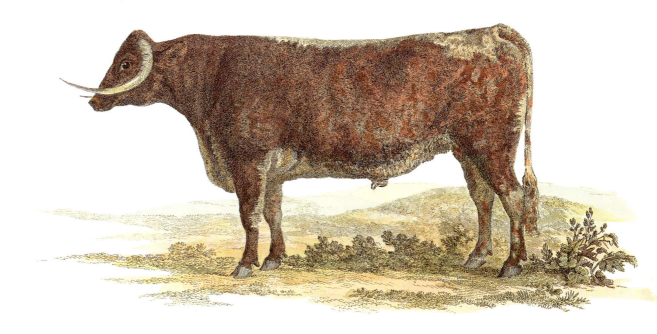

'A fat long-horned ox. Bred and fattened at Dishley farm in Leicestershire by Honeyborn, successor to the late celebrated Mr Bakewell'.
This ox was also sired by Shakespeare but its dam was Dandy (see pedigree above). The shoulder height of this ox was 4 ft 10½ ins (149 cm). The Longhorns pictured by Garrard are notable for having much higher rumps than shoulders. The height of this ox at the hind quarters was 5 ft 2½ ins (159 cm).

The Fat Long Horned Ox
- Shakespeare (bull b. 1778)
  - D (bull b.1772)
    - Twopenny
    - cow
      - Twopenny (bull b.1765)
        - Westmorland bull (Bakewell's original stock)
        - Old Comely (canley Cow from Webster)
      - Old Comely
  - Young Nell
    - Twopenny
    - cow from Mr Webster's stock owned by Fowler of Rollright, Oxfordshire
- Dandy

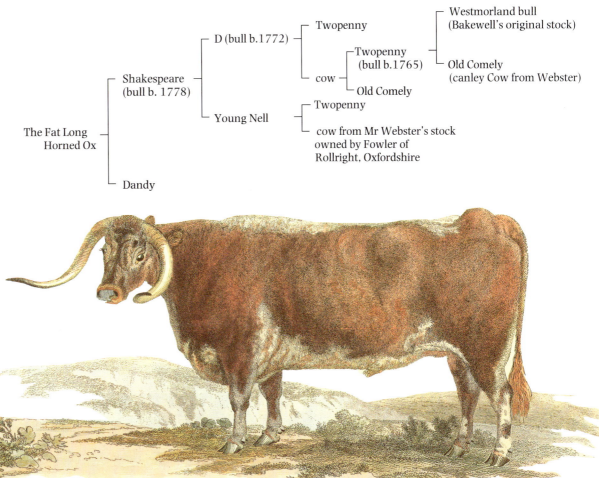

65

Their glory in Britain was shortlived and the Longhorn compared so unfavourably with the emergent Shorthorn (which was more fertile, more hardy and a better milker) that in a few decades, as Low recorded in 1842:

> . . . its reputation has passed away, even more quickly than it was acquired. It has given place to other breeds possessing characters as grazing stock, in which it is deficient . . . On the very farm on which Mr Bakewell's original experiments were instituted and completed, and within many miles around, there does not exist a single bull, cow, or steer of the breed which he had cultivated with so much labour.

Through the nineteenth century and into the twentieth a handful of enthusiasts kept the breed going, partly because the cattle are so decorative and partly because of their known ability to produce a good cross-bred animal. A Longhorn-Hereford heifer won the gold medal at Smithfield in 1847 as best female of any breed. In 1923 it was possible for Wallace & Watson to record that 'a good demand exists for bulls for crossing purposes, and for Longhorn females to form new herds'. Sixty years on, these remarks still hold good and, in fact, Longhorns are one of the most sought after rare breeds.

Nowadays many farmers use Longhorns for crossing, partly because heifers and cows served by Longhorn bulls tend to calve easily. The breed's tendency to lay down fat as a thick layer, rather than intermingled with the lean, could be a major advantage for breeders seeking to produce lean beef. Indeed one of the paintings which Professor Low commissioned from Shiels, and which today hangs in the Royal (Dick) School of Veterinary Studies in Edinburgh, depicts a Longhorn cow with a massive accumulation of fat on the rump, 'so great as to provide a kind of deformity in the fatted animal'.

Devon

No other British county has two native breeds of cattle. If the South Devon and the Devon had not fulfilled different farming requirements one would have died out or the two breeds would have merged, but this did not occur and there are striking contrasts between the histories of the two breeds and between their respective adaptations.

The Devon was certainly very well known to fashionable and progressive farmers in the later eighteenth century and the élite herd was that of Francis Quartly of Great Champson, Molland. These cattle were bred for beef and draught with milking qualities only a minor consideration. Quartly spent much money buying up good stock, building on the work of his father who is believed to have started breeding these cattle in 1776. When the first herd book was published in 1851 Quartly was acclaimed as the saviour of the local race and the founder of the new breed. Royalty, too, was interested in Devon cattle and the Duchy Home Farm at Stoke Climsland in Cornwall had a famous herd.

Reasons for this popularity among the aristocracy, in sharp contrast to the obscurity which shrouded the South Devon, are probably several. The use of Devon cattle for draught was well established, and for the progressive farmers of the time the use of oxen for work was not old-fashioned but was an important part of their agronomic manifesto. James Black, of Morden, Surrey, wrote this about his 'Devonshire ox team' in 1784:

> The Ox gives us his labour; the expense of his keep, compared with that of the
> Horse, is as one to two; and after his labour, when fat, he is worth two shillings per
> stone.
>
> The Horse is, as to his keep, as two to one, diminishes in value every year; at last his
> Skin is sold for five shillings.

Another possible reason is that of all the lowland breeds of England, the solid red breeds, that is the Devon and Sussex, are the ones for which a continental influence is least apparent. National as well as regional pride could have underlain the breed's popularity. Indeed even at the height of the Shorthorn's popularity (1860–69) there were no pedigree Shorthorn breeders in north Devon.

The high prices paid by Quartly in building up his herd show that good Devon cattle were well in demand even at that early stage. Throughout the nineteenth century this popularity endured and the 1908 cattle census recorded 454 694 Devon cattle. They made up 6.5 per cent of the national herd, and were, after the Shorthorn, the most numerous single breed, just beating the Ayrshire.

Also known as the Red Ruby, the Devon has always tended to be kept as a beef animal and there has been no programme in recent years for modernization (apart from the development of a polled strain using American blood). The Devon has always been acclaimed as one of the most attractive of the ancient lineages of locally-developed breeds of cattle in Britain and it is perhaps particularly worthy of conservation in its traditional form.

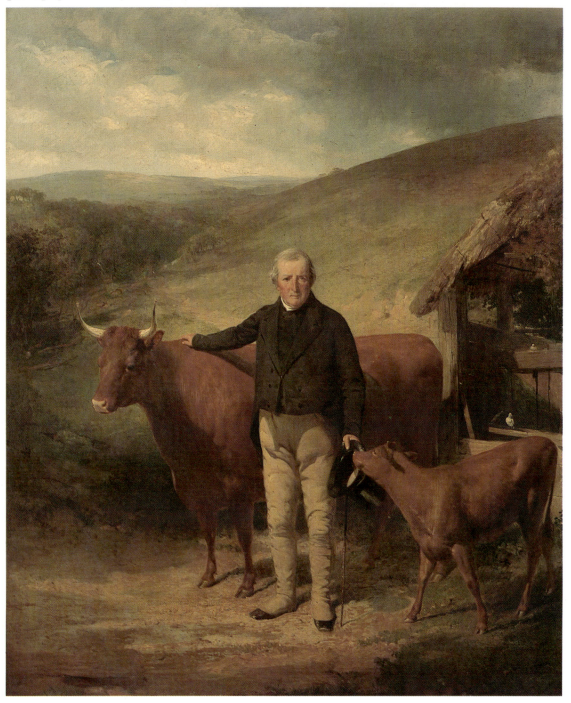

Portrait by Thomas Mogford (1800–1868) of Francis Quartly, whose name is firmly linked with the development of the Devon breed.

Thomas Coke (created Earl of Leicester in 1837) owned Holkham Hall in Norfolk from 1776 to 1842. 'Coke of Norfolk' (pronounced 'Cook') was responsible for major advances in the eighteenth century 'agricultural revolution'. His favourite cattle for milk, meat, and draught were Devons. Lithograph by W.H. Davis.

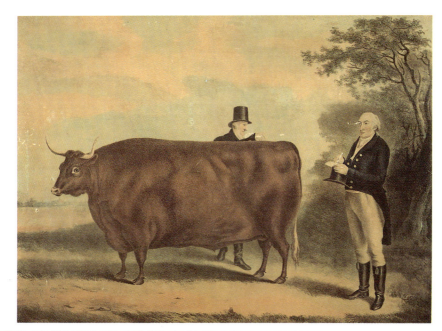

Evelyn Dunbar (1906–1960) painted *The Bail Bull* in 1945. This bull is running with a dairy herd on the Hampshire Downs. The bail is the mobile milking unit in the background. The Land Girl is tempting him closer with a bucket of feed so he can be led in for the night.

South Devon

Traditionally this breed has been thought of as arising from a cross between Devon cattle, which are supposed to have been found originally in the south as well as the north of the county, and Channel Island cattle, particularly the Guernsey. Fraser, reporting in 1794 to the Board of Agriculture, commented on the large size of mature oxen in Devon and the fairly high yield of very rich milk which was already noteworthy.

Often known as South Hams, these cattle were recognized as a distinct type by the end of the eighteenth century. In many ways they were treated as a 'poor relation' of the Devon, and Youatt (1835) wrote that 'their flesh was not so delicate as that of the North Devons. They do for the consumption of the Navy, but they will not suit the fastidious appetites of the inhabitants of Bath, and of the metropolis.' According to Punchard's Royal Agricultural Society of England report on the agriculture of the county in 1890, a Guernsey cow was kept with each ten or twelve South Devons to improve the milk. In the minds of pedigree cattle breeders the South Devon was firmly associated with the Guernsey. However there was enough interest in the breed for the herd book to be started in 1891, with 130 members in the breed society.

The 1908 cattle census showed the breed to number 96 991 cows, which was 1.4 per cent of the national herd. The South Devon made its first appearance at the London Dairy Show in 1910, and subsequently these cattle attracted comment at such events because of their size. Cows could weigh 16 cwt (800 kg) and bulls up to one and a half tons (1500 kg).

Squeezed on the one side by the Friesian and on the other by the colour-marking beef breeds, the South Devon was, after 1945, in the same predicament as the other dual-purpose breeds, but on 1 January 1948 South Devon milk was deemed eligible for a special premium on account of its high quality and within two or three years membership of the breed society which had been about 450 was doubled. However, the use of the breed as a dairy cow declined in the 1960s after withdrawal of the premium.

The South Devon has never been extolled as a prime beef breed and was not highly selected for early maturity or small size, which were the targets achieved so efficiently by the Hereford, Aberdeen-Angus, and Beef Shorthorn by the 1950s; to such a degree in fact that the field was wide open for the Charolais and the other continentals, where frame size had not been reduced. It is the only native British breed where the addition of foreign blood to increase size for beef production has been optional. It has also kept a reputation, probably due to its strong representation in South Africa, for being adaptable to hot climates, and this was one factor in the contract made in 1969 to export 700 South Devon bulls to the USA over the next five years; however, this contract was not carried out. The South Devon is now a beef breed and the breed society claims that its docility makes it very suitable for bull-beef rearing.

Double-muscled South Devons are also known. This condition, which is genetically determined, varies in expression from individual to individual and is particularly associated with the Belgian Blue and Charolais, but it also occurs in other breeds including, very occasionally, the Friesian. In double-muscled cattle, muscles are hypertrophied, that is the fibres proliferate and the muscle becomes much more bulky while the associated skeleton remains relatively undeveloped. The characteristic has always been widespread in the South Devon, and when the breed had a dairy role this was not particularly welcome as it was associated with difficult calvings and poor milk yields.

The South Devon is also unusual because although all other British mainland breeds of cattle only carry the gene for haemoglobin A, the South Devon has the gene for haemoglobin B as well. Both genes are also present in Jersey and Guernsey cattle and in more southerly breeds such as the Indian humped zebu. The island of Jersey was closed to the import of cattle in 1789, but it is recorded that zebu cattle were crossed with Channel Island and Devon cattle on the mainland around 1795 to 1805. Whether the haemoglobin B gene entered the South Devon by way of the Channel Islands breeds or the zebu is not known.

In this context it is worth quoting a footnote of Garrard (1800). It is a comment from Mr Parsons to Garrard:

> I shall have the pleasure of shewing you my new Devons, which as a painter, I
> know you will say have a finer claim to positive beauty than any you have yet seen
> – they are Calves got by an Indian Bull given me by His Grace the Duke of Bedford,
> upon two year old new Devon Heifers, and are as fat as quails at a month old . . .

A dehorned South Devon bullock belonging to Mr W. Burden of Detling, Kent; painted by Diane Rosher in 1984.

Sheeted Somerset

Sheeted Somerset painted by Shiels.

For 150 years, sheeted cattle were associated with Broadlands House in Hampshire and with the family of the Countess Mountbatten. Passing mention of the breed was made by the early surveyors of the west country and in letters and books on cattle from 1722 onwards. Lord David Stuart, writing in 1970, described how sheeted cattle resembled 'a red cow of North Devonshire or West Somersetshire, with a white sheet thrown over her barrel; her head, neck, shoulders and hind parts being uncovered'. They were 'quite well known, though perhaps not very common'.

The sheeted cattle were probably brought to Broadlands in 1736 and in later years it seems fresh blood for the herd came from belted Welsh cattle purchased at Bangor. However, the herd had to be destroyed 200 years later as so many of the cattle reacted positively to the tuberculin test.

Low (1842) described polled and horned varieties of the Sheeted Breed of Somersetshire, which he also indicated were becoming rare. Writers of the late nineteenth century described sheeted cattle in herds all over England and Wales but there is no strong evidence of connection with the Broadlands herd or with the breed described by Low.

The Broadlands cattle, and presumably the others of Sheeted Somerset type, were probably the result of a cross between Dutch cattle and local Devon cattle, with some Channel Islands influence. It is likely that pure-bred Devon and Channel Islands cattle were more attractive for the pedigree breeder who otherwise might have developed the type into a commercially viable breed.

Gloucester

Brown finch-back cattle were one of the identifiable local races of middle-horned cattle in the early eighteenth century, inhabiting Gloucestershire, South Wales, and adjacent parts of Somerset and Wiltshire. The early history of these cattle is almost unknown but they seem, like the Suffolk Dun, to have been recognized widely as adequate dairy animals which were replaced by new breeds without having been given a chance to show how they could improve under modern management. Some development of the breed had taken place but by whom, and when, is uncertain. These cattle were associated with the making of Double Gloucester cheese, the manufacture of which involves milk being allowed to 'settle' twice.

Elsewhere the up-and-coming Longhorn had no great reputation as a dairy cow, but Marshall (1789) reported that it was replacing the local breed in the dairies of Gloucestershire. Cheesemaking was a highly competitive industry throughout the nineteenth century, with local varieties gaining or dwindling in availability and popularity and with the

Gloucester cattle in the Wick Court herd of the Misses Dowdeswell in the late 1960s. The white back and rump are characteristic of the breed.

Glamorgan cattle were very similar to Gloucesters. A favourite of King George III who kept a herd at Windsor, Glamorgan cattle performed all the carting, rolling, and harrowing tasks in Windsor Park as well as providing milk. Extinction in the face of competition was predicted as early as 1824. Mr Edwin Bradley, who bred this five year old cow painted by Shiels in 1842 was a great Glamorgan enthusiast. Writing in 1842, he stated 'I certainly have had a struggle to redeem the merit of the breed which had almost become extinct through the neglect of our county farmers, who, during the period of our last protracted warfare were reduced to break up most of our rich and valuable pastures for the purpose of cultivating corn, which bore such an enormous price during those times.'
By 1900 the Glamorgan was extinct, a fate only narrowly avoided by the Gloucester.

importation of American and Canadian cheeses towards the end of the century. Liquid milk markets developed and this was where the dairy herds of Gloucestershire concentrated their effort, capitalizing on railway links to London. Shorthorns yielded more milk than the old breed, and had the Duke of Beaufort not enthusiastically adopted the Gloucester and set up his Badminton herd, the breed would have died out. This herd, which was kept to about 100 head, was run as a dairy unit, and Stout's monograph of 1980 records the ways in which it gave rise to other herds and was represented at the 1909 Royal Show at Gloucester. In 1919 the Gloucestershire Cattle Society was formed and the herd book started. Crosses were made with other breeds and much experimentation carried out, notably by Earl Bathurst who also used Shorthorn and Dynevor cattle for crossing purposes.

The 1923–4 foot-and-mouth epidemic struck particularly hard in the county and many small breeders lost their cattle. The Gloucester could not regain the impetus which it had acquired as a breed, partly because of the death of the Duke of Beaufort and partly because the breed could not compete commercially with other dairy breeds either in quality or quantity of milk; neither could it rapidly acquire a role as a dual-purpose animal.

In 1927 there were 177 cattle in 13 herds but nothing slowed the decline which led to the Society closing in 1950. At this time only fifty cattle remained, in two herds, Earl Bathurst's at

The Gloucester cow Blossom was celebrated because of her connection with Dr Edward Jenner of Berkeley, who established in 1796 that inoculation with cowpox could protect a person against smallpox. Sarah Nelmes, a milkmaid, caught cowpox from Blossom. Lymph from Sarah was transferred by Jenner to James Phipps, an eight year old boy. James did not develop smallpox when inoculated with smallpox matter – a historic finding and a vital step in the development of preventive vaccination. This painting is by Jenner's great-nephew Stephen Jenner.

Cirencester Park, and the herd of the Misses Dowdeswell at Wick Court. The former was heavily cross-bred, the Wick Court herd survived in isolation. The Cirencester herd was soon dispersed while the Wick Court dispersal sale in 1972 led to the reformation of the Gloucester Cattle Society which is now actively conserving the breed.

In January 1985 there were 209 registered animals, 40 to 50 not registered, and of the total, 134 pure-bred mature females and 23 bulls were used for pure breeding. There are only six herds with more than five animals and two-thirds of the total breed is within or near Gloucestershire.

Hereford

Probably the most famous county breed of cattle, the Hereford was the creation of tenant farmers who expected it to work in the yoke for five or six years before being sold to graziers for fattening for the London market. The material with which the early Hereford breeders had to work was a docile animal which could be fattened on grass. Prominent among these breeders were four generations of the Tomkins family whose endeavours may have begun as early as 1720, ending with the dispersal of their herd in 1859.

A Hereford cow from the Earl of Egremont's stock, engraved by Garrard. She had a shoulder height of 4 ft 3 ins (130 cm).

The breed was ideally suited for the new trade in store cattle for fattening near London, and it ousted the Longhorn from the west Midlands, as well as becoming a popular breed for gentleman farmers elsewhere. While there were famous grey Herefords, and mottle-faced varieties, the white face was becoming a Hereford hallmark before it was firmly fixed in the breed by John Hewer. This conferred a uniformity of type which, along with the already famous beefing qualities, more than compensated for most farmers for the lack (until 1846) of a herd book.

Herefords were well represented at the first Royal Show in 1839 where they had their own class; it was clear that the breed and the grass-fattening husbandry system it was a part of was soon to come into competition with the Shorthorn. At this time, however, even though Hereford cattle from the top herds were offered for sale with their pedigrees, many fashionable breeders were still attracted to the Shorthorn by the well-established herd book. The early

Also the property of the Earl of Egremont, this bull had a shoulder height of 4 ft 6 ins (137 cm).

maturing Shorthorn's adaptability to feeding in stall or yard was a great and overwhelming advantage for high farming, where the manure of the cattle, carted to the fields, was essential for the growth of the heavy cereal crops on which the system depended.

For the Hereford the answer was to reduce the age of fattening, and this enabled the breed to hold its own as a store beast. As a sire on the dairy herd, the Hereford was found early on to confer beef qualities on cross-bred calves, and the Hereford descent is always clear from the white face which all such progeny have, although so far as is known the early breeders did not have this in mind when they fixed the white face characteristic.

The élite market for bulls for export really became prominent around 1875 when many North American ranchers adopted the Hereford to upgrade their ranch herds which were of Spanish origin. The American Hereford Cattle Breeders Association restricted registration of British Herefords from 1883, and South America then became the main market. Exports were badly hit by outbreaks of foot-and-mouth in the 1920s and home sales for beef breeding by the depressed state of agriculture generally. At this time, the only real development was in the use of bulls on the home dairy herd.

In 1950 exports of bulls to the USA were resumed, but North American bred bulls were now providing very stiff competition for British Herefords. During the years of separation, the American Hereford had become a thoroughly researched and costed breed, for which produc-

Garrard (1800) wrote of this breed that 'the oxen are in great repute for purposes of husbandry, the plowing of the County of Hereford being almost wholly done by them'. The shoulder height of this ox (far left) was 5 ft (152 cm).

Garrard wrote about the heights of Hereford cattle that 'the bulls [are] generally from 13 to 14 hands; the cows about 13; oxen from 15 to 17 or 17.2 and 18; but 15 or 15.2 is the common height of working oxen' [one hand = 10.16 cm]. This fat ox (left) had a shoulder height of 5 ft 2¾ ins (159 cm).

The sale ring at Stocktonbury in 1884, with Thomas Carwardine's bull Lord Wilton, then eleven years old, coming under the hammer for a thousand pounds. This bull was very influential in British Herefords; Anxiety 4th, bred in the same herd, dominates the pedigrees of American Herefords.

About 7000 Hereford bulls are registered each year in Britain. Pure-breeding Herefords exist primarily to satisfy the market for cross-bred cattle rather than to produce beef. The breed has traditionally been used when dairy cows are to be inseminated by a beef bull, but this usage is declining dramatically from the level of the early 1980s when around 60 per cent of such inseminations were by the Hereford. This painting (right) by Susan Crawford (1981) shows the modern beef conformation and lack of horns, characteristic of many of today's Herefords.

tion data and economics could be presented. Furthermore, since 1902 the polling factor had been concentrated into a Polled Hereford Herd Book and the American Hereford was much better adapted than the British to the world beef market.

The Hereford has never given a very lean carcass, fat being distributed throughout the muscle, so the Charolais and the other continentals were better adapted to the home market for bulls. It was therefore necessary to introduce the modernized carcass of the American Hereford to British Herefords. Two methods were used; the polling factor was introduced by crossing with naturally polled breeds like the Galloway and Aberdeen-Angus, and the same result was achieved, with carcass improvement, by importing naturally polled stock from New Zealand, Canada and the USA. These imports began in 1955, and by 1985, 85 per cent of Hereford registrations in Britain were of the polled variety.

ANDROSSPOLL 1 PLUTO

Sussex

Three English lowland beef breeds had emerged in the eighteenth century; the Devon, Hereford, and Sussex. All were used as draught oxen, an application that was not compatible with the production of premium beef as such cattle needed well-developed forequarters to bear the weight of their load, but they lacked the rounding-out of the hindquarters expected of the best beef animals.

The use of oxen for draught was declining throughout the eighteenth century, but their steady pace and high tractive effort at low speeds were ideal for the heavy clays of Sussex, and indeed this county was their last stronghold in England. In spite of their suitability, the trend towards the use of horses for farm work was not to be reversed and the abandonment of oxen continued. Roads became better and new machines such as the reaper came in, which oxen could not pull quickly enough. Cheap feed grain was imported from overseas, and a strong market developed in the towns for work horses, so that the final arguments for the use of oxen collapsed and by 1913 only a dozen or so teams were at work in Sussex. The last team was probably one near Eastbourne which was disbanded in 1929.

The Dyke Road Mill stood at the back of the Dyke Road Hotel in Brighton. It was moved there from its previous position in Belle Vue Fields on 28 March 1797 by eighty-six Sussex oxen.

Sussex farmers managed their ox teams with a view to the practicalities of fattening cattle for sale. The oxen were not overworked, as it was believed that if they exerted their full strength they would not fatten so well. Teams usually numbered eight, or ten or twelve in 'stiff land', while horses performed the same work in teams of four. Sussex oxen were usually yoked in pairs. Once trained (by the age of two and a half to three years) they were put in the plough and worked gently so as not to retard their growth. They were slaughtered when six or seven years old.

The Sussex had as good a reputation as the Devon and the Hereford as a beef breed in the early 1800s but it declined in both numbers and in status during the next fifty years; while its rivals improved, the Sussex stood still. Boxall (1972) has documented its history. The Napoleonic Wars prompted the replacement of pasture by cereals and generally it was the small farmers of the High Weald who kept the breed going. Indeed, Boxall concluded that the breed would have died out had all farmers been free to follow the change to arable farming.

By the 1840s the standard of excellence was the quick-maturing Shorthorn, though the merits of the Devon and Hereford were also appreciated. The Sussex compared very unfavourably and it was only when its use for draught had become insignificant that selective breeding for beef could be started, with the 1860s and 1870s seeing the re-emergence of the breed on the national scene. Much of the credit seems due to Edward Cane of Berwick, in East Sussex. Another famous herd was the Petworth, which exhibited cattle at the first Sussex County Show in 1793 and which is still at the forefront of the breed.

As part of the revival, a herd book was started in 1879, foreshadowed in 1842 when a local farming newspaper began to publish pedigrees. In 1908, the Sussex, numbering only 19 660 cattle was still one of the least numerous of the main British breeds. In 1938 the situation was similar, with only 235 bulls licensed that year, compared with 23 730 Shorthorns. Today the breed has a strong reputation for siring good quality beef calves from the dairy herd, with few calving problems. In 1980 a Breed Development programme was started involving crossing with Limousins and the breed now has a following abroad, with around half a million in southern Africa.

Like every other breed, Sussex cattle have always had their enthusiasts. Rudyard Kipling had 'only six and twenty' Sussex cattle but he cherished, like so many other cattle breeders, the dream of breeding a superlative bull:

> He shall mate with block-square virgins –
> kings shall seek his like in vain,
> While I multiply his stock a thousandfold,
> Till an hungry world extol me,
> builder of a lofty strain
> That turns one standard ton at two years old!

Garrard (1800) wrote of this Sussex heifer: 'allowed in all respects to be the handsomest Fat Beast that had ever trod the pavements of that market' [Smithfield].

Petworth General 26th was teamed with the cow Brede Gipsy 2nd as the Sussex entry for the Burke Trophy which they won at the Royal Show 1984. General was owned by Messrs B.W. Kemsley & Son and was painted by Diane Rosher in 1984.

Red Poll

The Suffolk Dun, a polled breed which at the end of the eighteenth century had a very high reputation for dairy merit, was combined with two Norfolk breeds (one horned, the other polled) to produce the Norfolk and Suffolk Red Polled Cattle. The Suffolk was dun, red, red and white, or brindled; the Norfolk breeds were red, often with a white face. The breeds were merged in 1846 following the amalgamation of the counties' Agricultural Societies. In 1862 there were classes at the Royal Show and in 1874 Henry Euren, helped by the Revd George Gilbert, published privately the first volume of the herd book. Also involved was a veterinary surgeon and farmer of Holt, Norfolk, the grandfather of the famous animal scientist John Hammond FRS.

When Arthur Young referred to the counties' cattle in 1794 he commented on the haphazard nature of their breeding and said that there were no bulls older than three years; perhaps Henry Euren had this criticism in mind when he arranged the foundation stocks named in his herd book into twenty-four 'groups'. Each group bore the name of a locality or of a famous breeder and was divided into tribes, each consisting of all descendants, male and female, of each ancestral cow. This provided the means of working out pedigrees quickly and relatively easily.

The original distinction between the two county stocks, the Suffolk herds for dairying and the Norfolk for beef, had practically vanished by 1900, and by this time horns (only seen as loose structures about one and a half inches (4 cm) long, developing at about twelve months of age) were seldom, if ever, seen. Where the polled condition arose remains unknown, although Bailey & Culley in 1794 attributed it to a cross with a Galloway. The origin of the milking qualities of the Suffolk is also unknown. Some link with Dutch cattle is likely but has not been demonstrated. Breeders have always been proud of the Red Poll's milking performance; they first published milk records in their Herd Book in 1892. Lord Rothschild kept Red Polls at Tring and in 1903 published comparisons among this breed, Jerseys and Shorthorns. In quantity and quality of milk Red Polls were placed between the other two.

In 1908 only about four out of every thousand adult cattle in Britain were Red Polls, and major expansion of the breed was taking place in the USA rather than in Britain. Red Polls did, however, increase numerically up to 1945. There was never a grading system and breed expansion was the result of pedigree cows being kept breeding for longer. The breed also spread out of East Anglia, and by the start of World War II was found all over England south of the Mersey and the Wash. After 1945 the Red Poll became rather fashionable and, apparently, animals that should have been culled were sold for breeding. Disillusionment with the results is a possible explanation for the crash in numbers which took place in the 1950s.

To get a quick improvement in milk yields the breed society decided to cross with Danish

'Starling of the true Norfolk breed, in the 36th year of her age, the property of Charles Money Rainham.' Native Norfolk cattle were scorned by modern agriculturalists but were perfectly adequate for local needs. Arthur Young described the breed as possessing 'no qualities sufficient to make it an object of particular attention'. However, judicious crossbreeding in the context of local selection procedures led to a valuable breed, the Red Poll, which has had considerable influence at home and abroad. Painted by an unknown artist.

Prize Heifer 2 Years and 8 Months Sept.r 21st 1841.

In 1846 the Norfolk and Suffolk breeds of cattle were formally amalgamated. This painting shows how before that date the two breeds had been merging to produce cattle very like the modern Red Poll.

84

Left: 'A Suffolk Ox. Bred by Mr Brett of Burnham near Holkham, Norfolk.' This ox had a shoulder height of 5ft 6 ins (168 cm) and a weight of 271 stones 6 lb.

Right: 'A Suffolk Cow. In the posession of Samuel Whitbread Esq.'

Historians, notably Trow-Smith (1959), have speculated that had the Suffolk Dun survived as a distinct dairy breed it could have assumed the role which the British Friesian adopted. While in the early 1900s a few herds were still being distinguished as Suffolk cattle (notably that of Prince Duleep Singh, near Thetford) the process of combining them with Norfolk cattle was almost complete by that time.

Red cattle, and 6 bulls and 87 heifers were imported in September 1961. The importation diminished the beefing qualities of the breed, and also cost the breed society dear in terms of resignations of some important breeders. The scheme failed and the Danish Red section of the Herd Book has dwindled, with no registrations in 1985.

In 1986 there were twenty herds of Red Polls, thirteen of which were milked, the rest being run as suckler herds. In size they ranged from 120 cows down to 22. Abroad, the breed has done well in Canada, North and South America, Australia and New Zealand, and in the West Indies.

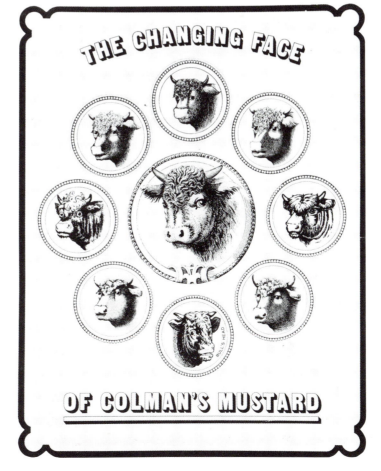

THE CHANGING FACE OF COLMAN'S MUSTARD

All that remains of the Norfolk breed is the stuffed bull's head mounted in Colman's mustard shop in Norwich. This was the model for the Colman's trademark. There is no relic of the Suffolk Dun apart from pictures and photographs.

5   Black and white cattle

The new husbandry methods of the nineteenth century did not succeed in meeting the entire demand for beef, so in this sense Bakewell's aim was not realized. Imports of live cattle were vital and came in vast numbers from Europe and the New World, beginning in earnest in the late 1850s after the repeal of tariffs. After 1876 only cattle for immediate slaughter were admitted and about 70 000 were imported annually. The trade came to an end with the banning of all live imports in 1892 (foreshadowed by a temporary ban, because of disease risk, from 1887 to 1889). Chilled and then frozen meat imports took over.

A high proportion of the live imports were cows and it seems that many of these were sold to the town dairies and after completing one lactation were fattened for beef. These dairies were disease ridden and virtually unregulated and contributed to the appalling levels of child mortality of Victorian times. Several factors combined to finish off town dairying in the 1870s. Notable among these were outbreaks of contagious pleuro-pneumonia, which killed 187 000 British cattle in 1860, and rinderpest, which was introduced to London after a twenty year absence on a Russian cattle boat in 1865 and wiped out 500 000 cattle in two years. Economic reasons included development of the railway network, and the decision of many lowland farmers, as a result of the agricultural depression of 1875, to change from arable or mixed to dairy farming.

Cattle from the Netherlands had always been well known as milk cattle, and by 1900 there were probably thirty or forty herds of Dutch cattle in Britain, known as Holsteins, Frieslands, or Friesians. In the 1908 cattle census, there were still too few of the new breed to merit enumeration. The ban on import of live cattle for breeding prevented the import of bulls and superior cows for breed development.

At this time the Shorthorn, though in theory dual-purpose, had become first and foremost a beef breed. The Ayrshire was a very high yielding dairy cow with very little beef value. The rather scarce Guernsey and Jersey catered for the wealthy end of the milk market. The Suffolk Dun had vanished, and its descendant, the Red Poll, was not a particularly high yielder. There was clearly scope for the further development of the Friesian, which had already secured some important enthusiasts. The Strutt family at Terling, Essex, had, since the depression started in 1875, been dairying on the lands vacated by their tenants who had abandoned farming. The Hon. Edward Strutt was the pioneer who shared the scientific attitude of his brother Lord Rayleigh, winner of the Nobel Prize for Physics in 1904. Strutt had no interest in pedigree breeding but he began milk-recording in 1896, and he clearly showed the superiority of the Dutch cattle.

James Bateman painted *Commotion in the Cattle Ring* at Banbury Market in 1935. Dairy bulls like this Friesian are well known to be more dangerous, or at least more unpredictable, than bulls of beef breeds.

The British Holstein Cattle Society was formed in 1909 when the first volume of the herd book was produced. 'Holstein' was changed to 'Friesian' in 1918. The foundation stock had been influenced by several sources including Canadian Holsteins which had exerted a strong influence on Clydeside herds. Cattle were registered after inspection and by 1913, 1000 males and 6000 females had been approved.

In 1914, 39 bulls and 20 cows were imported, having been selected in the Netherlands from good, milk-recorded stock. This importation was highly successful. In 1922, 28 bulls and 55 cows were imported from South Africa, but the expected increase in milk yields did not result. Further Society importations were made in 1936 and 1950 from the Netherlands and in 1946 from Canada. The dangers of dependence on a small number of sires were demonstrated when of the thirteen bulls imported in the latter batch, three were found to carry lethal recessive genes. A less serious deviation from the ideal genotype was the presence, in five of the bulls, of the recessive gene for red colour. Breeders reluctant to discard good milking cows because they were red and white rather than black and white formed the Red and White Friesian Cattle Society in 1951; in 1976 this Society imported stock from the Netherlands. In 1985 the Society merged with the British Friesian Cattle Society.

Kenneth Ogborne painted Moneymore Bunty 6 in 1983, the year in which she was a supreme champion at the Royal Dublin Spring Show. The spread of the Friesian was slower in Ireland than in the UK, mainly because the importance of the trade in 'Irish stores' (part-Shorthorn fattening animals) was such that reduced efficiency of dairy production was accepted by the Irish Government during the 1950s and 1960s in the interests of protecting Irish breeders of Shorthorns.

During and after World War II dairy farming was vigorously encouraged. In 1950 there were 40 per cent more dairy cattle than in 1914, and the British Friesian was now the most successful dairy breed, dependent upon large inputs of concentrates for a large output of milk. The other breeds had retained importance only where local conditions gave them a particular advantage – the Shorthorn in north-west England and south Wales, the Channel Island breeds in the south of England, and Ayrshires in most of Scotland.

Canadian blood was imported by the Society in 1946, the intention being the improvement of udder conformation and an increase in milk and butterfat yields. The 220 animals imported differed markedly from the traditional British Friesian in carcass conformation and indicated that a higher priority was being placed on milking rather than beefing qualities. Since the 1950 Society importation from the Netherlands only private importations have taken place. There were numerous private importations from Canada in the 1970s, which peaked in 1973 when 749 cattle were brought in.

Today the British Friesian makes up 85 per cent of the British dairy herd. It is not only the most important breed for milk, but also the most important single breed for beef. The breed has increasingly been taking on the attributes of the Canadian Holstein and it is becoming rather rare to encounter a cow bearing a pedigree showing that her ancestry is due entirely to stock imported from the Netherlands. Many records including milk yield and sale price continue to be set by British Friesians. One such record, which illustrates the genetic consequences of modern reproductive technology, was set in 1983 when Grove Speculator, an eleven year old bull belonging to the Milk Marketing Board and kept at their AI centre at Ruthin in North Wales, had fathered 80 000 calves by artificial insemination.

British Friesian heifers. Each animal in this picture would be expected to produce 5.7 tonnes of milk, nine times her own body weight, in a lactation.

6 Channel Island cattle

In the late eighteenth century cattle from the Channel Islands were known collectively as Alderneys, taking their name from the nearest of the Islands to England. Imports of these cattle had taken place at least since 1724, and by 1775 more than 900 were coming in annually, two thirds of them from Jersey. From 1789 the importation of live cattle into Jersey from France was greatly restricted (from 1819 in the case of Guernsey) and since then the island herds have been closed and the export of breeding stock has been a major part of the island economy.

Channel Island cattle soon became famous and widely dispersed. They were crossed with, among others, Devon, Ayrshire and Shorthorn cattle; perhaps the greater size of the Guernsey indicates some admixture of Shorthorn blood. The richness of Channel Island milk was always renowned and Culley stated in 1807 that 'the Alderney breed is only to be met with about the seats of our nobility and gentry, upon account of their giving exceeding rich milk, to support the luxury of the tea-table'. Youatt in 1835 still included cattle imported from Normandy under the name of Alderney, although he knew that these were larger animals than the diminutive and, in his opinion, very ugly Channel Islands cattle.

These cattle were for a long time not strictly distinguished by their island of origin, although in 1834 the newly formed Royal Jersey Agricultural and Horticultural Society drew up a scale of points relating specifically to the Jersey. In this they were inspired by Col. le Couteur and Mr Michael Fowler, the latter the best known importer of Jersey cattle to England. The first classes for Channel Island cattle were held at the Royal Show, Southampton, in 1844. The Herd Book for the island was established in 1866 and the English Jersey Cattle Society was formed in 1878, by which time the annual importation of Jersey cattle was running at about 2000. This was a heavy drain on the island's stocks, and led to exports being restricted in the 1880s.

In the mid nineteenth century there was a fashionable demand for silver grey Jerseys, led by Mr Philip Dauncey's herd at Horwood, Buckinghamshire, which was founded in 1825 and dispersed in 1879. Following this, whole colours (that is, solid or self-colouring as opposed to broken or mixed) were favoured. Fashionable involvement like this led to a greater public awareness of the breed's dairy qualities as rich owners were prepared to show their stock, and at the London Dairy Show in 1879 253 Jerseys were entered, this being the single most numerous breed. It was in 1871 that separate classes for Jersey and Guernsey at the Royal Show were finally adopted as standard practice.

Guernsey cattle are larger than Jerseys and characteristically have broken colouring. They have never been quite so subject to the dictates of fashion. The relative numerical importance of the two breeds depends critically on pricing policies for milk and at present the Guernsey is not favoured by these policies.

In 1875 only a third of the milk produced was on the liquid market. By 1908 liquid milk had become established as part of the national diet, and the census of that year recorded 101 233 Jersey and Guernsey animals, 1.4 per cent of British cattle. From about 1920 the sale of milk in bottles grew rapidly and this drew attention to the cream line, favouring the Channel Island breeds with their high butterfat and attractive colour.

By 1955 Guernseys comprised 5.3 per cent of the milk cattle of England and Wales (at 130 400 cows) and Jerseys 2.6 per cent, the most numerous breeds after Friesian, Shorthorn and Ayrshire. World-wide there are about six million Jerseys; they are the second most numerous dairy breed, the Holstein or Friesian occupying first place and the Brown Swiss third.

In both Jersey and Guernsey there has always been a wide range of colours, even including black with white. It is striking how in this painting, made by W.A. Clark in the 1920s, of Guernsey livestock, three of the seven cattle are pied. A Golden Guernsey goat is included in the group. It is believed this picture was copied by Clark from an older work.

This cow was painted by F. Babbage in 1910 and considered to be 'the most perfect Jersey cow ever seen'. She won first prizes in 1904 and 1905, was imported into England by Mr A. Miller-Hallett and won numerous further prizes from 1906 to 1908. The cow was finally exported to North America where she was sold by auction for £1500. Channel Island cattle tend to have a southern distribution and are often described as less hardy than other breeds, but this has never been formally tested.

Carcass dissections have shown that Friesians have more subcutaneous fat and this could have an insulating effect. This difference is an expression of the fact that the Jersey is a more extreme dairy type cow than the Friesian. The different fat deposits in a cow are drawn upon in different ways during lactation, and the subcutaneous fat is not so readily mobilized. The fat deposited within the abdomen is more accessible and this is where Jersey cattle store their fat. It is also interesting that although a Jersey cow is only 60 per cent of the weight of the average Friesian, the two breeds produce almost the same total butterfat (around 200kg) in a lactation.

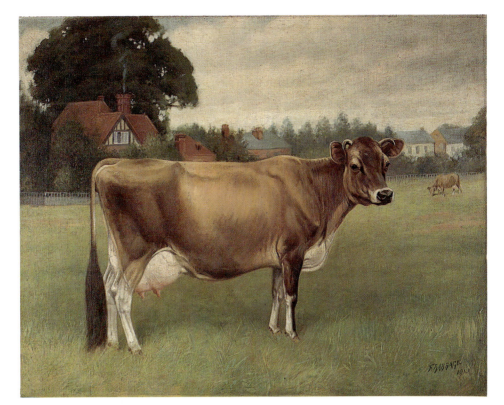

James Ward RA (1769–1859) painted this life-size portrait of Alderney cattle in 1820–22. It measures 3m × 4.8m, and was an emulation of the picture of a bull by Paul Potter, the Dutch master. The cattle depicted were the property of Mr Allnutt of Clapham.

Edwin Douglas (1848–1914) painted *Alderneys (Mother and Daughter)* in 1887. The initials on the tree trunk are those of the artist and his wife. Jersey cattle are unusual in that their calves are particularly small, even taking into account the small size of the cow, and as a result calving problems are rare.

7 Continental beef breeds

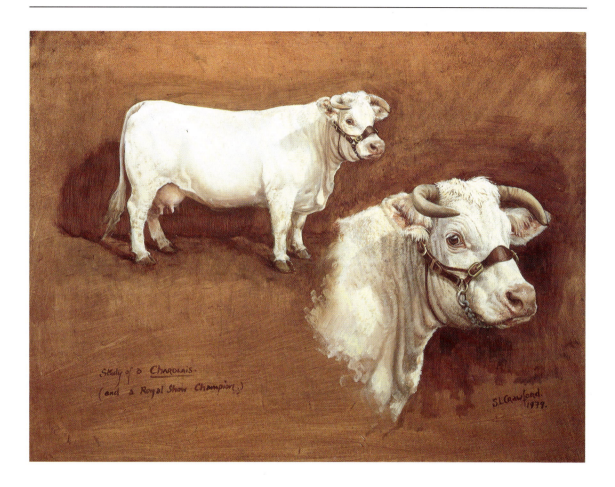

Study of a Charolais by Susan Crawford.

The continental breeds have played an important part in the process whereby beef production in Britain has increased from 550 000 tonnes in 1946 to 1 million tonnes in 1983; currently over 90 per cent of the beef eaten in Britain is home produced. In 1960 the Terrington Report recommended the carefully supervised introduction of the Charolais to Britain. Semen from 27 bulls became available from 1 March 1962 and official tests were conducted, as a result of which the breed was adopted. There was spirited resistance from the British breed societies to the proposed importation because the Charolais had already made inroads into the export of British pedigree bulls; in 1960, 163 bulls and 522 females had been exported from France, mainly to South America. The Limousin and Simmental followed in 1970 from France, and Germany and Switzerland respectively.

Most British beef comes from the dairy herd; the figures are 43 per cent from steers and heifers bred in the dairy herd, 19 per cent from culled dairy cows, 34 per cent from specialist

Royston Reuban

'The Charolais cattle we mentioned in our notice of the Paris Agricultural Show are very handsome, but half a century behind our improved breeds, and therefore of no use here; the Royal Agricultural Show was not instituted to give prizes for objects of natural history.' This report in the London Illustrated News in 1856 is a striking example of how it is often the relatively unimproved breeds that have been shown, subsequently, to fill a new agricultural need which improved breeds are too refined to satisfy. Susan Crawford painted this Charolais cow (right), a Royal Show champion called Cigale, from the de Crespigny herd in Sussex in 1975. The bull, Royston Reuban, also a Royal show champion, is from the Bassingbourn herd in Hertfordshire and was painted by Kenneth Ogborne in 1984.

Mr C.B. Playle and his daughter Mrs C. Parker, prominent breeders of Charolais cattle in Hertfordshire, with trophies.

The distinctive characters of the Charolais are easily recognized in this bull and cow of 1856, figured by Baudement in 1861.

The Limousin as it was in 1856, figured by Baudement in 1861.

beef herds and 4 per cent from Irish store cattle. The British beef industry therefore is intimately associated with the dairy herd of 3.2 million dairy cows, 2.6 million of which are of Friesian type. About 36 per cent of British beef cows are Hereford-Friesians, or other crosses of Friesian cows with beef bulls. These cows are got in calf by a terminal sire to produce the slaughter generation. Within recent years the Charolais has been the single most popular terminal sire for the beef herd, siring about 40 per cent of the calves produced for slaughter while the other continental beef breeds have a 10 per cent share. However, the Charolais figure may now be on the decline. The Limousin is the most popular continental beef breed sire for dairy cattle in England and Wales.

Beef cattle sired by bulls of traditional British breeds are ready for slaughter at a relatively light weight. If they are kept on the farm after this time their subsequent weight increase is in the form of unwanted fat. Cattle sired by continental bulls are later maturing; they do not lay down excess fat until a later age, so carcass weight is greater than that of cattle sired by bulls of British breeds. The choice of terminal sire breed is highly relevant to the amount of beef produced. At the time of the Limousin-Simmental importation, it was calculated that if these breeds eroded the demand for Herefords as a terminal sire by 100 000 inseminations the effect would be to increase the supply of home produced beef by 7000 tonnes, through the larger size of the slaughtered animals.

8   Parkland and primitive sheep

The native sheep of the Scottish islands received little attention from the early improvers of livestock even though the softness and lightness of Shetland wool had long been renowned. Youatt (1837) quoted a proposal that Shetland sheep should be crossed with Merinos but it is unlikely that this occurred to any marked extent. The wool of most Shetland sheep in his time had to be plucked as it was shed rather than shorn, so that only a very small weight of wool was collected from each animal. Even today, if Shetland sheep are not shorn they will shed their wool.

Included here as primitives are the Soay, Orkney (North Ronaldsay), and Shetland. These are small, coloured sheep with short tails and a single pair of horns. In addition there is the Castlemilk Moorit, a relatively 'new' breed created by the Buchanan-Jardine family in Dumfriesshire at the beginning of this century, which resembles the Soay. Then there are the four-horned breeds; the Hebridean (previously called the St. Kilda) and the Jacob, which have long been kept as park animals in England, and the Manx Loghtan, native to the Isle of Man.

The primitive sheep are relics of unimproved breeds that have probably remained much as they were when first introduced to the British Isles between the Neolithic and the Iron Age periods, some 4000 to 2000 years ago. Soay sheep represent a link between the mouflon of the Mediterranean (a relic of an earlier stage in sheep husbandry) and the relatively unimproved breeds of north-west Europe. The mouflon is a dark, chestnut brown sheep with a light saddle, light belly, distinct rump patch, dark ruff and short, broad tail. In contrast to this uniformity Soays of many different colour patterns exist.

The four-horned sheep have close affinities with those in Iceland and Scandinavia, and they could well be descended from sheep brought from these countries many times over since the Viking Age, one thousand years ago.

The Jacob, or Piebald sheep as they were called until recently, are of unknown origin. Some authors have claimed that they are from Syria or Spain, others that they came from Scandinavia. The relatively long tails of these sheep indicate that they have been interbred with improved stock and they are no longer a primitive breed.

The primitive and four-horned sheep first came into the public eye at an exhibition arranged by John Elwes FRS at the Royal Show held at Bristol in 1913. Elwes's intention was to promote the crossing of these sheep with improved breeds for the purpose of scientific study and he was in advance of his time in his mention of the 'Mendelian Theory of Breeding' in relation to his experiments.

Soay

Prehistoric people probably introduced the sheep whose descendants run feral today on the islands of Soay and Hirta, St. Kilda.

For hundreds of years a population of around 200 Soay sheep has inhabited the island of Soay, being the property of the laird (Macleod of Macleod). They were described by Martin, whose account of a visit to St. Kilda was published in 1698. In 1930, after the people of Hirta were evacuated and all their livestock removed or destroyed, 107 Soay sheep were moved from Soay to Hirta. Since then the Hirta population, like that of Soay, has lived in a feral state.

On Hirta nearly all the males and about half the females are horned. About 5 per cent are self-coloured, that is, they are the same colour (black, or dark or light brown) all over. The other 95 per cent have a colour pattern like that of the mouflon but not usually so strongly developed. About a quarter of them have a pale buff brown body with the mouflon pattern muted; the rest have a dark brown body. About 5 per cent of sheep have white spots or patches, and some are piebald.

Several ornamental flocks of Soays have been started over the last hundred years from direct importations from the island of Soay, and selective breeding has been practised on them. Some of the owners believed that the variety of colours in the flocks of St. Kilda indicated a departure from the concept of the 'true Soay'. Darwin had stressed that variations among individuals in an animal population are an invariable result of human intervention and clearly it was felt that the Soay flock must have been influenced by sheep carried across to Soay from Hirta in early days, or that this variability of colour was a result of domestication. The evidence so far assembled (from test matings planned in accordance with what is known of the genetics of sheep coat colour) suggests that white patches, piebald patterns, and all the

General Pitt Rivers, the pioneering archaeologist, appreciated how bones of the Soay sheep he kept in his own park were very similar to those found in archaeological sites. His sketch made in 1888 shows a horned ram, a horned ewe, and a polled ewe.

colour anomalies seen on Hirta are part of the normal colour range and are probably not due to influence from improved sheep in recent times. However, from the turn of the century until today breeders have held different views about the 'true Soay'. For instance, the Woburn flock of the Duke of Bedford (founded in 1910) was selectively bred following the premise that the dark type was 'correct' and that all ewes should be horned, whereas Lord Lilford preferred light brown sheep. In common with the Duke of Bedford, and, indeed with all owners of park Soays, so far as is known, he wanted the ewes to be horned.

In 1963 a group of sheep was removed from Hirta, comprising a selection of colours, and ewes with or without horns. Flocks derived from this group have been kept separate from other flocks, and represent lines derived purely from Hirta without selection for one or other type. This means that the breed is now divisible into Park Soays and Hirta Soays. Some other islands now have populations of these sheep. Stock from Woburn was used to found the flocks on the Welsh islands of Skokholm (1934) and Cardigan Island (1944).

Soays are attractive sheep which are comparatively easy to look after and today many are kept as pets, although they are also finding commercial applications. In the last fifteen years many farmers have been developing husbandry systems which involve high stocking rates (possible with small ewes like Soays) and the production of good quality lamb by using terminal sires like Southdowns or Ryelands.

Soay sheep on Hirta, St. Kilda.

Orkney and Shetland

Of all the British sheep, it was the archaic stock of the Scottish highlands and southern uplands that was most completely effaced by modern agriculturalists. By 1840 the native breeds predominated only on Orkney, Shetland, the Uists, Barra and Benbecula. On the mainland, no attempts had been made to improve the native breeds (although it is probable that a southern Scottish race gave rise to the Cheviot). In the central and west Highlands native sheep were rare; in the east and north-east they were replaced altogether or were graded up to Blackface. The last flock of native mainland sheep is stated by Trow-Smith (1959) to have died out near Inverness in 1880.

These native sheep had not been managed in a way which could be considered as commercial, having been part of a peasant subsistence economy. Sheep were housed at night for the sake of their manure; and the lamb competed with the family for the ewe's milk. Today's native sheep of Orkney and Shetland represent relics of these pre-1800 sheep, as well as a tangible link with the ancient rural economy.

Native sheep have long since disappeared from most of the Orkney Islands, except from North Ronaldsay, and elsewhere in Orkney where there may be a few flocks in which native blood predominates. On North Ronaldsay, the old system of runrig or strip cultivation was abolished late, most of the island being squared off into little farms in 1832. A wall was built around the island, either at that time or earlier, and the native sheep have been kept on the foreshore ever since. There are probably around 3500 there now.

The major part of the diet of these sheep is seaweed, but ewes are brought in to grass fields before lambing in May. The sheep are adapted in several ways to eating seaweed, preferring the red seaweeds which are of highest nutritional value (comparable to oats or ley grasses). Seaweeds are low in copper and these sheep are so efficient at extracting it that when kept on normal pasture they have been known to absorb enough copper from the herbage to develop copper poisoning.

North Ronaldsay sheep are very variable in colour, making it almost impossible to pick out two that are identical. They can be grouped into a range of colours, white and greys being most numerous, but also black or dark brown, and reddish brown. There are hairy and woolly individuals. Almost all males, and about 20 per cent of females, bear horns (in Soay sheep, on Hirta, almost all males and 46 per cent of females bear horns).

This is at present the only rare breed of sheep which has an established physiological specialization, namely its adaptation to a diet of seaweed. This remarkable attribute is one reason why the first action, in 1973, of the newly formed Rare Breeds Survival Trust was to buy the small island of Linga Holm in Orkney in order to establish a reserve flock on it,

Shiels has depicted here a pure-bred Orkney ram and ewe, and the produce of a cross with a Cheviot ram. In Shetland, many breeds and notably the Cheviot have been crossed with the native sheep to improve carcass weight.

because the single North Ronaldsay population was too much at risk from disease or oil spill.

In Shetland today the native breed is firmly associated with the famous knitwear and woollen industry, the only case in Britain, apart from Channel Island milk, where a breed name is also seen as a trade mark. The Shetland is the only British breed which can produce wool approaching that of the Merino in its fineness. Shetland sheep exist in a very wide range of colours, but white predominates now in the islands and it is the rarer colours such as moorit (tawny red-brown) and markings such as katmoget (dark underparts with light upper parts) that need to be conserved. Registrations of sheep on the British mainland, which tend to be of the rarer colours and markings, are handled by the Rare Breeds Survival Trust, while the Shetland Flock Book Society (founded in 1927) caters for the island sheep. Crosses with Cheviots are particularly frequent in the islands and these result in successful meat-sheep.

In earlier days only a handful of Shetland sheep were found outside the islands. For instance, of the 88 343 Shetland sheep qualifying for the Hill Sheep Subsidy in Scotland in 1941, 12 were in Aberdeen, 113 in Inverness, 150 in Orkney, 24 in Ross and Cromarty, and 88 044 in Shetland. Now it is one of the most numerous rare breeds and it could well have a bright commercial future, as the high stocking rates on marginal land possible with small sheep like these mean that they could produce abundant meat-lambs, particularly when crossed with a Down ram. On the hill a live weight of around 22 kg is normal for a ewe; under parkland conditions ewes and rams can weigh up to 45 kg and 65 kg respectively.

On North Ronaldsay, Orkney, the native sheep graze on growing seaweed or eat the weed cast up by the tide.

Illustrated here are commercial Shetland sheep in their native islands and two pet ewes kept for their wool by Mrs J. Arnold in Hertfordshire. A primitive characteristic is that these sheep tend to shed their fleece in June or July; rooing, or plucking, is the traditional way to remove the loose wool, although today most Shetland sheep are clipped.

Jacob

The origins of the multi-horned, piebald breed which in Britain is now known as the Jacob have fascinated several generations of livestock historians. Sheep farmers have been interested in the breed too, mainly because of its high fecundity but recently because of the value of its wool to the hand-spinner. Elwes reported on his crossing trials at the Royal Show in 1913, finding the 'Piebald' or 'Spotted' breed to fatten well on rather poor grazing, to 'produce more doubles than any breed which I have kept', and to have excellent mothering qualities as well.

Mr Heatley Noble, a contemporary of Elwes, investigated the history of the breed, concluding that originally the sheep were two-horned and some northern, perhaps 'Hebridean' cross introduced the second pair. Elwes (1913) agreed with Noble, pointing out that this is the only long-tailed breed of sheep where the four-horned character is common (a long tail is a feature of improved British breeds, while four horns are noted particularly in the primitive sheep of the western British seaboard).

Whether or not this, the most economical explanation of the origin of Jacob sheep, is completely correct may never be known. Certainly these sheep were widespread at least from the mid eighteenth century, each flock being associated with some country house. In many families, tradition led to some romantic or exotic origin being assigned to the flock. Such names as Barbary, Jacob, Spanish, Persian, African, Zulu and Syrian were used. The Charlecote flock (still in existence, in Warwickshire) has been credited, by the evidence of a letter dated 1756, with a Portuguese origin, but it is possible that the exotic sheep were added to a pre-existing flock. Flocks of four-horned piebald sheep were known at Wentworth (Yorkshire) and Tabley (Cheshire) at about this time and the Portuguese sheep apparently brought to Charlecote are unlikely to have given rise to all these other flocks.

Robert Bakewell knew about these ornamental sheep, which were clearly a combination of rather different types. On 23 December 1791 he wrote thus to Arthur Young:

> Yesterday a man forward for Bradford Hall with an Iceland Ram . . . The Ram I send
> to you to be made any use of you think proper but I believe the sort as little inclined
> to get fat and eat as much food as most I have yet seen. I want an ewe of the same
> kind if you can procure them for me they must be of the four-horned kind, these are
> some of nearly the same colour called Spanish but differ from these, those I can
> have near home. . . . (Pawson, 1957)

This two-horned Jacob ewe and her lamb were painted by Denis Curry in 1984.

Unlike so many other basically ornamental breeds, the Jacob sheep never fell on bad times. The Jacob Sheep Society was founded in 1969 and the next year there were 116 members with 3000 sheep. The breed has increased markedly in numbers and in 1986 78 tonnes of wool described as Jacob were marketed (representing the clip from over 30 000 sheep, and probably including contributions from many other coloured breeds).

Hebridean, Manx Loghtan, and Castlemilk Moorit

Many English parks had held flocks of 'St. Kilda' sheep for a long time when Elwes wrote the Guide for his display of primitive sheep and their crosses at the Royal Show in 1913. Elwes designated these sheep 'Hebridean' and considered they were descended from some primitive breed of the outer Scottish islands, perhaps with a Norwegian connection, and mixed to some degree with the Scottish Blackface.

Various accounts published in the nineteenth century refer to multi-horned sheep in the Hebrides and it seems likely that visitors to the islands, wanting to bring sheep of this kind home for ornamental purposes, chose multi-horned animals. Elwes reported that 'most, if not all' of the flocks in England had been crossed with small black sheep of 'Welsh, Breton or other breeds'. Notable flocks were kept at Woburn by the Duke of Bedford and at Ascott, near Leighton Buzzard, by Mr Leopold de Rothschild.

Four-horned Hebridean sheep at the Cotswold Farm Park, Gloucestershire. Two-horned Hebrideans are rather more common and the four-horned character is associated with the congenital 'split eyelid' defect. Four-horned Hebridean rams are difficult to keep as when they fight they can easily split their skulls open. For this reason many of the four-horned sheep of this breed on display in farm parks, including these ones, have been castrated, reducing their aggressiveness.

In 1976, 900 000 of these 1p coins bearing a picture of a Manx Loghtan sheep superimposed on a map of the island were struck at the Pobjoy Mint, Sutton, for the Isle of Man.

Surrounded with curlicues, a Manx Loghtan ram decorates the 5p coin.

Manx Loghtan ram on the 10p coin. The 5p and 10p coins were struck in 1982.

Hebrideans are black or dark brown, commonly becoming grey in old age. In 1973 there were about fifteen owners and around 400 breeding females, but today there are over 800 ewes in about sixty flocks.

Similar sheep are associated with the Isle of Man. Elwes reported that some of the native sheep were white, others black, but the loaghton or mouse-brown colour was the most favoured. No pure-bred Manx sheep was polled. The multi-horned character was more prevalent in these sheep than in any other breed he knew, 'possibly owing to selection'.

As early as 1913 the breed was practically extinct; its previous history is not recorded although it is presumably from the same stock as the Hebridean. Unlike the Hebridean it received very little interest from mainland owners, but was conserved in its native area by local people. Notable among these has been Mr Jack Quine who kept the stock going in sufficient numbers for flocks to be built up on the Isle of Man by the Curraghs Wildlife Park and the National Trust. Today's Manx Loghtan population is descended from these, and the pedigrees of many mainland sheep trace through the flock of four rams and three ewes which came to the Royal Agricultural Society of England from the Whipsnade collection.

On the mainland the breed has increased greatly in numbers since the Rare Breeds Survival Trust initiated registration and gave encouragement to breeders. As with the Hebridean, a surprisingly high proportion of the sheep registered as foundation stock in the early years of the Trust continue to be represented in today's lamb crops, indicating that genetic conservation has been substantially achieved.

One of the least known and, by its originators, least publicized breeds of sheep created this century is the Castlemilk Moorit, developed by the Buchanan-Jardine family to beautify their park, near Lockerbie in Dumfriesshire. The flock, numbering a hundred sheep, was put up for sale in 1970 and one ram and ten ewes were bought for breeding. The rest were slaughtered. Mr Joe Henson, then collecting stock for the Cotswold Farm Park, bought the major portion. He described them as 'coloured like blonde Soays with short Soay type tails and a very distinct Mouflon pattern. The ewes all had fine upward and outward curving horns . . .'

Manx Loghtan, Soay, and possibly Mouflon sheep are the likely origin of the breed, which was always known locally as Moorit Shetland, implying some early influence of the latter breed, although there is no resemblance now. In 1983 the breed was accepted by the Rare Breeds Survival Trust and now there are eight flocks totalling 120 breeding ewes.

Castlemilk Moorit sheep.

9   Sheep of hill and mountain

When Culley wrote his book on livestock in 1807 there appear to have been few sheep in the Scottish Highlands and what there were he described as the 'dun-faced breed'. This was a primitive short-tailed sheep with short wool and it was probably at least in part ancestral to the Cheviot on the Scottish Borders, for at the end of the eighteenth century, according to Ryder (1983) the Cheviot still had a dun face and the rams were horned. Culley, however, claimed (incorrectly) that the dun-faced breed was hornless and, like the Herdwick and many other breeds of British sheep, it was reputed to be descended from Spanish sheep rescued from the Armada. The history of the mountain breeds is therefore very unclear and it is also unresolved as to how the Scottish Blackface, a breed which according to Trow-Smith (1959) originated on the Pennine hills, adapted so well to the Highlands of Scotland. Although Culley did not describe the Blackface, this breed was already widespread in Scotland at the beginning of the nineteenth century where it was known as the Linton, and was already being differentiated into the Blackface, the Swaledale, and a number of regional types. Today a flock of feral sheep inhabits the St. Kildan island of Boreray and this is clearly a relic of the nineteenth century type of Blackface.

To the south, in Cumberland, the Herdwick breed of the Lake District appears to have changed very little since the time of Culley who described this sheep as, 'a lively little animal, well adapted to seek their food amongst these rocky mountains'.

Culley made no mention of the Welsh sheep although as Youatt (1837) stated, 'sheep farming is the staple business of the [Welsh] agriculturalist'. During the eighteenth century every part of the Welsh uplands had its own variety of mountain sheep. They were small, unimproved animals, some with white faces and some dark and they still held the primitive trait of shedding the fleece. By the early nineteenth century the sheep of the Welsh mountains were being crossed with improved breeds, and factories for the weaving of wool were becoming established. Before this, all spinning and weaving was carried out at home and the people of Wales were noted for wearing a great quantity of woollen clothes. Youatt, quoting Luccock (1809), commented that, 'In the north, females are generally seen in felt hats, and large blue cloaks; those of the south are fond of exhibiting the various coloured borders of two or three petticoats, and wear upon their shoulders a square piece of red cloth'.

The hills of south-west England carried similar sheep to the unimproved Welsh. They were white-faced and horned and were known as the West Country Mountain sort. These gave rise to the Exmoor Horn, the Dartmoor, and the Devon Closewool, and further north on the southern Pennines to the Penistone, Silverdale, and Whitefaced Woodland.

Cheviot and North Country Cheviot

The Cheviot seems to have been developed from a local race of the north Northumberland – Berwickshire region, with some input from Lincoln rams around 1760. Robson of Bowmont Water has been credited with this successful crossing. Later on, some Leicester blood was introduced. The new breed is said to have grown a fleece of fine wool but the infusion of longwool blood increased the quantity while diminishing the quality of the wool. This was the usual penalty of a Leicester-type cross, but on the credit side there was an improvement in meat conformation and speed of maturity.

The changes made in the fleece of the sheep were relevant to its subsequent commercial career because in the areas which it was to colonize over the next century, a fleece with a proportion of hair in it (as in the Blackface) casts off the rain more efficiently than one which is of purer and more valuable wool.

In 1792 Sir John Sinclair, who gave these sheep their name, took a flock of five hundred to Caithness. His aim was to prove that the Cheviot could thrive in the far north. Indeed, by this time there was a good deal of competition between Blackface and Cheviot, the latter being favoured on the lower, grassy hills of the Scottish border country.

In the north the breed was extensively developed and Merino blood introduced. The result was a large animal with comparatively good wool, and in 1912 the North Country Cheviot Sheep Society was formed. The Cheviot of the south country, the native area, is smaller and more blocky, and has remarkably erect ears unlike those of the North Country Cheviot which come out at angles of 45 degrees from the head. The North Country Cheviot is hornless, while horned rams are occasionally met with in the south country.

Shiels' painting of a Cheviot ewe and lamb, bred by Mr Thomson of Attonburn, Roxburgh.

North Country Cheviots in
Sutherland and Caithness,
photographed by Glyn Satterley.

North Country Cheviots in Sutherland and Caithness, photographed by Glyn Satterley.

In the south of Scotland too there were many developments. James Brydon of Eskdalemuir produced an 'Improved Cheviot', with good conformation, rather small size, but reduced hardiness which was catastrophically demonstrated in the winter of 1859–60 when very many of these sheep died. The Lockerbie type, in the south-west of Scotland, tended to have small horns. The North Country Cheviot is a bigger, more placid and apparently more profitable sheep which is why on many occasions Border sheep farmers have attempted to replace their Cheviot flocks with 'Northies'; this happened during the 1920s when scrapie wiped out many Cheviot flocks on the eastern borders. It soon became clear that the incomer was less hardy than the longer-established type and the latter was reinstated on many hill farms.

A distinct variant is the Brecknock Hill Cheviot which is also called the Sennybridge. It seems that the McTurk family who moved to Brecknock, in Wales, in 1856 brought Cheviots and that previous, undocumented introductions had been made some fifteen years earlier. The present day Brecknock Hill Cheviot has some Welsh Mountain blood, which has given extra length to the body, but which has not improved fleece quality. Rams are occasionally horned. About 1 per cent of the Welsh wool clip today is described as Cheviot.

Once the Cheviot had achieved identity as a breed, late in the eighteenth century, crosses were made with the Culley brothers' New Leicesters and the resulting cross-bred found favour as a meat animal. This was the start of the famous 'stratification' of sheep breeding which became fully developed between 1870 and 1939. In this period, cheap grain imports led to much arable land being turned over to grass and the arable sheep, namely the long-woolled breeds and to some degree the Down breeds, declined. The new arrangement was for cross-bred ewes to be bought in and mated with suitable 'meat sires' or 'terminal sires' to produce the slaughter generation. Cross-breds combined the fecundity of the crossing sire (traditionally the Border Leicester, so far as the Cheviot and Blackface were concerned) with the hardiness and mothering ability of the upland ewe. For the lowland flockmaster where shepherding operations often coincided with arable commitments, purchase of ewes made considerable sense in reducing shepherding effort. There were also fewer disease problems as the hill ewes used for the production of cross-breds came from closed and relatively disease-free flocks.

The 'Scotch Halfbred' (the Border Leicester – Cheviot cross), has always been noted as a dam of fat lamb, especially on very good land where the cross-bred derived from the North Country Cheviot is particularly favoured. Indeed, flocks of 'pure Halfbreds' became numerous in southern Scotland between 1870 and 1900, involving the use of Halfbred rams on Halfbred ewes, but the lambs were heavy and slow-growing and over the next two decades the use of Halfbred rams as the terminal sire was discontinued. Other flocks of Halfbred ewes were crossed with Border Leicester rams to give 'three-parts-bred' lambs but these were too fat for the twentieth century market and Oxford, later Suffolk, rams were used instead, which remains the general pattern today.

From the *Sheep Sketchbook* by Henry Moore. Scotch Halfbred sheep on a Hertfordshire pasture, drawn with a ballpoint pen.

Herdwick

Taylor Longmire of Ambleside (1841–1914) painted many watercolours of Lakeland scenes and livestock. These Herdwicks were probably painted in the 1860s.

White-faced, short-woolled sheep are native to the west of England and, in parts of their range, they have come into competition with the black-faced heath sheep of the Linton type. In the north particularly, the white-faced sheep have lost the competition; for instance the Silverdale was extinct by 1920. Other breeds remain only as relics, the scarcest of which is the Whitefaced Woodland, the only rare breed of hill sheep. Certain Pennine breeds represent crosses between the black-faced and white-faced types, for example the Derbyshire Gritstone (also partly of Leicester descent) and the Lonk.

The Herdwick is another such hill sheep, white-faced although the colour changes with age. It is noted for its hardiness. In 1807 Culley had described the breed as a polled sheep with a few black spots on the face and legs, and with a fleece that had been coarsened by crosses with the Linton (the progenitor of the Scottish Blackface), the Swaledale, and the other black-faced mountain breeds. Indeed a decade earlier, in their survey of Cumberland, Bailey & Culley had attributed the white on the face of the Linton to Herdwick crosses. It seems highly likely therefore that the Linton may be responsible for the hairiness of the fleece of the present-day Herdwick.

As early as 1844, a society was formed to support the breed, the West Cumberland Fell Dales Association for the Improvement of the Herdwick Breed of Sheep, and in 1855 classes for the Herdwick were included in the Royal Show at Carlisle.

Canon Rawnsley, the vicar of Wray who helped to found the National Trust in 1895, also appreciated the importance of the Herdwick as the native sheep of the Lake District and was involved in a sheep breeders' association founded in 1899. The breed was further strengthened by the formation of the Herdwick Sheep Breeders' Association in 1916, and the first Flock Book was published in 1920, when 2000 rams were recorded. After Beatrix Potter (Mrs Heelis) who was a childhood friend of Canon Rawnsley bought Troutbeck Park Farm in 1923, her attachment to Herdwick sheep grew very strong and she contributed generously both in her own lifetime and by bequest to the landholdings of the National Trust in the Lake District. The Trust now has over 87 farms there and about 25 000 Herdwick sheep which are leased to the Trust's farm tenants. Indeed, the keeping of Herdwicks on many of these properties was a stipulation of Mrs Heelis' will.

On the right is Mrs Heelis, better known as Beatrix Potter, the major Lakeland benefactor of the National Trust and a firm friend of the Herdwick breed.

Today the wool of the Herdwick is being promoted by the National Trust and in 1986, 147 000 kg were produced in Britain (compared with 2 090 000 kg of Swaledale and 78 000 kg of Jacob). Herdwick wool is one of the coarsest and is mostly used for carpets. The National Trust is also promoting its use for knitwear, and the Herdwick may well follow the Shetland sheep in becoming a breed whose name is synonymous with a retail product.

Herdwicks are particularly striking in that their lambs, by whatever ram, are dark brown or black, gradually developing their characteristic blue-grey colour later.

Herdwick sheep near Wastwater, photographed by Glyn Satterley.

Scottish Blackface

This Blackface ewe painted by Shiels has fine twin lambs. Traditionally the breed has not been seen as very fertile under free-ranging conditions, but extra feeding, if correctly timed, enables the Blackface ewe to fulfil her potential for producing and rearing lambs. 'Twenty five years ago', according to the Hill Farming Research Organisation in 1979, 'the few flockmasters who did practise some feeding were loath to admit it and reluctant to discuss it', on the grounds that extra food would reduce the hardiness of the breed. Today the extremely beneficial effect of extra food at certain times of year on flock fertility is widely accepted. An increase in lambing percentage (lambs born per hundred ewes put to the ram) from the figure characteristic of many flocks – 80 – to the 160 of which the Blackface is capable, is of enormous economic value.

It is impossible to trace their origin, there being no tradition of the sheep here being ever of a different kind.

Thus wrote James Naismyth, in 1795, of the black-faced horned moorland sheep of the uplands of Strathclyde. These sheep were also known as Lintons, after the market at West Linton where they were sold.

Even in Naismyth's day different kinds of black-faced moorland sheep were known, among them the Galloway sheep with a shorter but finer fleece and more white on the face. Today several races are recognized; the Lanark, Newton Stewart, Perth, and Lewis, all differing particularly in fleece characteristics but also in meat conformation. The Lewis is small and lean, while the Lanark is the largest. The Perth and the Lanark have the heaviest fleeces. The Newton Stewart is thought to have been the most improved, but it is no less hardy than the others. Within the breed there are three types of wool; Mattress, staple length 8 inches (20 cm), Medium, 6 inches (15 cm), and Short Fine, 3 to 5 inches (8 to 13 cm), with differences in the thickness and softness of the wool. Today the Blackface produces nearly half of Scotland's wool, for uses ranging from carpet manufacture to the weaving of Harris tweed.

The Blackface definitely came from south of the Border but details of its progress north are unknown, and the sheep which it superseded, called the Dunface or the Old Scottish Short-wool, is also something of a mystery. Relics of this race presumably endowed the Newton Stewart and the Lewis Blackfaces with the relatively fine fleeces they possess today.

Blackface ewes and lambs in Glencoe, Argyll, painted by Eirene Hunter in 1986. Underlying this peaceful scene is a complex social organization. Like all wild and feral sheep the Scottish Blackface has a strong female home range tendency. Hill flocks are hefted ('heft' describes both the group of sheep and the area to which they restrict themselves, or are restricted by shepherding) and this has its origin in the bond between the ewe lamb and her mother. On a given farm some hefts are usually better than others in terms of lamb production and this is a reflection of differences in grazing quality. When hill farms change hands the hefted flocks usually stay on the farm.

The husbandry system of the south-west, based on the Old Scottish Short-wool, was completely swept away when the Blackface took over. The process was repeated, starting around 1752, when the new sheep invaded the Scottish Highlands. In that year John Campbell took Blackface sheep from Ayrshire to Dunbartonshire. The Blackface was very rapidly established on the central and west Highlands because previously the emphasis had been on cattle and the hills were only pastured in summer. It took longer for the eastern Grampians and north-east uplands to be taken over as there were large numbers of native sheep to be replaced but by 1840 the native breeds and their husbandry were only to be found in Orkney, Shetland, and some of the Outer Hebrides.

At the same time, rivalry between the Cheviot and the Blackface was intense. Initially the Cheviot gained ground as its better wool found a ready market when the Napoleonic Wars cut off the supply of fine Spanish wool. By 1860 Cheviots made up nearly half of the hill sheep of Scotland, with the Blackface strong on the higher and harsher hills. The severe winter early that year did far greater damage to Cheviot flocks than to the Blackfaces, especially in the southern uplands, and the Blackface staged a resurgence there. Cheviots retained their hold on the country north of Inverness and were at their most widespread in 1870. However, after this high point Cheviots declined as cheap fine wool was imported and severe winters in the 1870s accelerated the process. The meat market also favoured the Blackface; lambs of this breed were more easily finished for the growing demand for high quality lamb.

Throughout its history, the Blackface has kept its reputation as one of the hardiest British breeds and it was in furtherance of this attribute that its promoters refused to have anything to do with pedigree breeding, flock books or a breed society until 1901. They felt that fashion points could easily become paramount, to the detriment of the breed. Even so, on many occasions particular rams or flocks have commanded enormous prices. Whether this has been the result of fashion or of a genuine commercial requirement is debatable.

There are very many factors to be borne in mind when selecting a hill ram. He must be able to pass on good characteristics of feet and teeth, fleece and meat conformation. Unlike the Cheviot, the Blackface is expected to cope with a tough diet including heather, and to obtain enough food long distances must be covered, usually over difficult country. With between five and ten per cent of the income of the typical hill farmer coming from wool, the most remunerative fleece must be bred for, that will protect the sheep against the weather. Since half of his lambs will be male, to be castrated and sold for meat, the ram must also be able to endow them with good fleshing characteristics. Perhaps therefore it is not surprising that the right Blackface ram can be worth over 40 000 guineas to his purchaser. Such prices are paid for rams that are going to be used in élite flocks which themselves supply rams to the commercial hill farms.

The husbandry system of Blackface flocks shows very many local variations, developed over decades in response to local conditions. The general pattern, however, is for old ewes to be sold (drafted) from the hill farms with their tough herbage to lush grassland often very far from their native hills. This is a valuable trade for the hill farmer. These drafted sheep are the main source of the cross-bred ewes which are currently the most widespread lowland sheep.

Boreray is one of the islands of the St. Kilda group, lying about 6.4 km off Hirta and Soay, which is where the Soay sheep are found today. When the people of St. Kilda were evacuated in 1930 all their husbanded sheep were removed except for the stock hefted on Boreray, which is inaccessible for most of the year. These husbanded sheep were of Blackface type and were a relic of the early days of the breed in the Outer Hebrides, when they were crossed with the local sheep, known as the Dunface or Old Scottish Short-woolled sheep.

On account of its isolation for fifty years, the Boreray is now a distinct breed. There are probably around 400 on the island. This photograph shows how both sexes are horned and that many have dark collars and tan markings.

Three ewes and three rams were taken off Boreray in 1971 to form a research flock at the Animal Breeding Research Organisation, Edinburgh. In 1981 the breed was recognized by the Rare Breeds Survival Trust. There are now around fifty on the British mainland.

Swaledale and relatives

The northern short-tailed, black-faced heath breeds were for most of the nineteenth century considered as one general type, and while early attempts were made to improve them with New Leicester crosses it was accepted that the most appropriate force shaping the breed was the natural environment. In Scotland these processes had led by 1900 to distinct types of Blackface; in the north of England, these local types became dignified with the status of separate breeds.

The early history of the different local races is hardly known, and the events and developments preceding the formation of the different breed societies have not been scrutinized by historians. It is known that the breeders of black-faced sheep in the north of England formed the Swaledale Sheep Breeders' Association in 1920 because large numbers of Scottish Blackface sheep were being sold south to the Skipton—Settle area and the Cleveland Hills; the local breeders countered this expansion and the twentieth century history of the northern black-faced breeds has been the repelling of the Scottish Blackface by the Swaledale, and an accelerating trend for Swaledale rams to be used in Scottish Blackface flocks to increase milking performance. Today there are just under 2.5 million Scottish Blackface ewes and about 500 000 Swaledales in Britain.

The Rough Fell and Dalesbred are the other particularly hardy black-faced breeds of northern England. The Lonk (of Lancashire) and the Derbyshire Gritstone are associated with

Swaledale by Alan Stones of Penrith (1986). This lithograph shows the grey or white muzzle and solid black face of this breed. Other features distinguishing it from its main rival, the Scottish Blackface, are more important commercially. The Blackface has always been considered better than the Swaledale as a meat-sheep, being blockier, shorter in the leg, and with a better developed shoulder. The Swaledale ewe, however, has the reputation of providing more milk for her lamb and being more active and therefore a better forager.

Sheepshearing in Baldersdale, by John Gilroy (1899–1985). These are Swaledales. The wool clipped from the hill ewe is only of small value in terms of British farming but the coat is crucial to the animal's winter survival, and the early Swaledale enthusiasts stressed the importance of a tightly binding undercoat which did not blow up and expose the skin to the wind. Many otherwise good sheep did not meet this 'coat and waistcoat' requirement and the Dalesbred became established as a distinct breed in 1930. Superficially the two breeds are very similar but the latter has a longer outer coat than the Swaledale. It also appears to have been outcrossed, possibly to the Scottish Blackface, at an early stage of its development as a breed, and its conformation, rather better than that of the Swaledale, seems to be due to these crosses.

rather less rigorous conditions. Both received local recognition at quite an early stage, the flock books being founded in 1905 and 1906 respectively. They are bigger sheep than the Scottish Blackface and the Swaledale and have closer, finer, heavier fleeces which seem to get waterlogged in very heavy rain, a fleece type they might owe to white-faced horned breeds like the extinct Silverdale or the Whitefaced Woodland or Penistone. The Derbyshire Gritstone is a polled breed and many rams are sold to introduce the polling gene to other hill flocks.

Different black-faced breeds are associated with different areas in the north of England. In the northern Pennines most flocks are Swaledale with some Dalesbred – further south, about 50 per cent are Dalesbred and 15 per cent Swaledale, the rest being Derbyshire Gritstone, Lonk, Rough Fell, and some cross-breds. Around the industrial areas the hill flocks are Derbyshire Gritstones, Lonk, Scottish Blackface, and Whitefaced Woodland. On the North Yorkshire Moors, Swaledales predominate in the west and south and Scottish Blackfaces elsewhere. Swaledales predominate in the Lake District though the Herdwicks are more widely regarded as characteristic. Derbyshire Gritstones and Lonks are strongly associated with Derbyshire and Lancashire.

Just as from an early stage the Border Leicester was extensively crossed with the Scottish Blackface and the Cheviot (as a crossing sire for the drafted hill ewe) so too the Wensleydale and later the Teeswater became associated with the Swaledale and its relatives. This cross produced Masham cross-breds; today, most drafted English hill sheep are crossed with

The Rough Fell (also called the Kendal Rough) has rather a localized distribution, being strongly associated with the hills to the north-west of Sedbergh. Traditionally these sheep are brought down to winter on lower ground. The Flock Book was started in 1926. The Rough Fell, with its irregular face markings and much stronger and longer staple of wool than that of the Swaledale in some ways resembles the Scottish Blackface. Taylor Longmire painted this sheep some time in the late nineteenth century; the view is looking south down Windermere, from Troutbeck.

Right: also painted near Troutbeck but a hundred years later. Portrait of Mule ewes and their Suffolk cross lambs, by Sylvia Frattini.

Bluefaced Leicesters to give the Mule. In his prize essay of 1849 on the adaptations of sheep to specific localities, Rowlandson described how hill ewes were mated with Leicesters or Southdowns, and then kept over the winter on sheltered ground feeding on turnips. Their cross-bred lambs were weaned at three months, and kept for three years before slaughter for mutton.

Today the black-faced hill breeds are basic to the British sheep industry, but 150 years ago their presence was barely acknowledged by the pedigree breeders who accepted, in Rowlandson's words, that 'only particular species of sheep are adapted to subsist on the dreary wastes where such breeds are found'.

Welsh hill breeds

The small, hairy-woolled native sheep of the Welsh mountains, unlike those of Scotland, were not swept away by the new black-faced or Linton breeds. Since the Neolithic period, sheep had been the dominant livestock of upland Wales, while in Scotland cattle had this role. The native sheep of Wales were hardy and numerous; hills were not bare of sheep, as they were in Scotland, and husbandry of sheep was a well understood part of the Welsh economy. A gentle transition had taken place in the eighteenth century from a cattle droving trade to one of sheep. Welsh mutton, from sheep fattened by Midlands graziers, was very highly regarded.

The early surveyors distinguished many local races of Welsh hill sheep and tended to regard them with contempt, because they were so very different from their ideal which was the big-bodied, fast growing lowland sheep with a heavy fleece. Indeed, on the lower lands where turnip cultivation was possible, Cotswold and Leicester sheep replaced the native breeds. Elsewhere, new breeds emerged from a combination of native and introduced sheep. On the highest ground the native sheep remained largely unaltered, unlike in the Scottish highlands where the local breeds were replaced by imported sheep. While the Cheviot established itself around Sennybridge, on good pastures, the Blackface made slight progress, becoming popular only around the Flint–Denbigh border.

'Breed of the higher Welsh Mountains. Ewe, from the high range of Glamorgan, north of Neath . . . Ram of the same race, in a state of improvement, bred by The Right Hon. Viscount Adare M.P. Dunraven Castle, Glamorganshire. Ewe, of the same descent, improved, bred by Viscount Adare.'
Breed development as illustrated here was only possible in areas where the grazing had been improved.

The Welsh and Scottish mountain breeds differ biologically as well as in their economic and social backgrounds. The former are much more varied in type and this has resulted in such offshoots as the Black Welsh Mountain, Torddu, Badger-faced Welsh Mountain, Torwen, and Balwen becoming more or less established as breeds in their own right. In Welsh Mountain sheep the gene for black colour is recessive to that for white, which explains why occasional black animals are seen in flocks of white Welsh Mountain sheep. The Black Welsh Mountain seems, however, to have a dominant gene for black colour.

Other important differences include aspects of fecundity; the Welsh Mountain is less prolific than the Scottish Blackface but has better survival of the lambs. It has a much less valuable fleece than the black-faced hill breeds; 42 per cent of Welsh Mountain and Welsh cross wool has coloured fibres in it which are not wanted. Of fleeces from the Scottish Blackface and the Swaledale, 25 per cent and 14 per cent respectively are downgraded for this fault. As a small bodied sheep the Welsh Mountain can be stocked at high densities and ewes whose breeding careers on the hills are over are usually freely available to lowland farmers while the equivalent Blackface ewes can be quite expensive.

This Welsh Mountain ewe lost her own lamb and was given Kitty Fisher, twin to Lucy Lockett, to foster. Kitty is a Kerry Hill cross. In the words of the artist, Denis Curry (February 1988): 'Kitty Fisher, now a handsome full-grown ewe . . . has just had her own fine lamb'.

Selling the Ram won the 1986 National Eisteddfod photography competition. It was taken at the South Wales Mountain Sheep ram sale at Nelson, mid Glamorgan, by Martin Roberts. These sheep are sometimes known as Nelson or Glamorgan Welsh sheep and are the largest bodied Welsh breed. Their breed society was established in 1948 and was revived in 1977 by which time there were twenty-six flocks with 13 501 ewes in the Society.

Traditionally, Welsh Mountain sheep only graze the mountains from April to October. Much of this land is common land and most is not productive enough to keep ewes over the winter. In some favoured areas, however, not all are withdrawn for the winter. Store lambs are sold in autumn and nowadays the old ewes are usually mated with Border Leicester or Bluefaced Leicester rams to give Welsh Halfbreds and Welsh Mules respectively. These cross-breds are popular on lowland farms in England as well as in Wales because their small size permits a high stocking density; in comparison with the Welsh Mountain, they are much easier to keep because they do not have such a proclivity to escape through imperfect fences.

Some of the development work which brought the Welsh Halfbred into prominence was carried out in the 1940s on the Cambridge University Farm. At that time there were no recognized channels for the sale of these cross-breds. Welsh Mountain ewes had often been crossed with Kerry Hill, Clun Forest, Wiltshire Horn, and Down breeds but few of these progeny were ever retained for breeding. An association was formed to promote the Welsh Halfbred in 1956.

The Welsh Mountain sheep is sometimes known as the Whiteface Welsh, to distinguish it from the Beulah Speckled Face (or Eppynt Hill) and Welsh Hill Speckled, which together account for 8 per cent of British hill ewes. These have arisen in ways which are almost completely undocumented, leading to relatively docile sheep which are larger and can capitalize on better vegetation than can their mountain counterparts. The Beulah is also the breed recommended by the Nature Conservancy Council for use by nature reserve managements in controlling scrubby growth on grasslands of conservation importance; it is freely available, relatively easy to keep, hardy yet placid, and seems to prefer tougher vegetation, unlike cross-breds like the Mule. Beulah rams are sometimes used to improve wool and increase size in mountain flocks. The Lleyn is another development of the Welsh Mountain which is becoming widely known because of its prolificacy.

The Welsh Mountain Sheep Society was formed in 1905; the Black Welsh Mountain Sheep Society in 1922. In 1984 Professor J.B. Owen reported that the Welsh Mountain is probably now Britain's most numerous breed of sheep. There are about 14 million breeding ewes in Britain and half of them are hill and mountain sheep. Of these about 2.5 million are Welsh Mountain, with slightly fewer Scottish Blackface.

The Lleyn of north-west Wales probably arose from a New Leicester – Welsh Mountain cross in the early nineteenth century though the improvement might have begun earlier with the use of Roscommon rams from Ireland. These sheep were abundant in northern Gwynedd (Caernavonshire and Anglesey) but from about 1940 numbers of pure-bred Lleyns declined and by the mid 1960s there were only about ten flocks with 500 ewes. A breed society was formed in 1970. The resurgence since then has been quite extraordinary with 231 registered flocks and about 4000 breeding ewes in 1985. The Lleyn is being vigorously developed using a Group Breeding Scheme. This involves participating farmers contributing highly prolific, milky and early-breeding ewes to a nucleus flock, where rams are bred for use by the participants. Such systems were first used in the 1960s by New Zealand sheep farmers but are, even today, uncommon in British sheep. This picture is of a pedigree ewe and her four lambs from Flock no. 247, Ty Coch Farm, Gwent.

Whitefaced Woodland

In the eighteenth century, white-faced, horned, close-woolled sheep were kept on the southern Pennines and on their western slopes as far north as the Kendal – Carnforth area. As the century progressed these sheep came under considerable pressure from the hardier black-faced heath types, which were better adapted to rain and to rough grazing. In the lowland parts of their range, other types of farming competed. As a result only relics of this type of sheep remained, the Silverdale (or Limestone) in the north and the Woodland and Penistone (around the valley and town of the same names, respectively) in the south.

The southern races seem to have contributed to the Derbyshire Gritstone and Lonk breeds, judging by the fact that these hill breeds are rather less hardy, but have a better fleece, than the other black-faced hill breeds such as the Swaledale. The Silverdale was acknowledged by the early surveyors to be better than the local black-faced sheep but it became extinct by 1920. As the nineteenth century progressed, Cheviot and Blackface crosses were made on the Woodland/Penistone. Presumably different degrees of crossing with these breeds are the origin of the differences which present-day breeders maintain exist between 'Penistone' and 'Woodland' sheep.

Merino rams are known to have been used in the flock of the Duke of Devonshire at Chatsworth, Derbyshire in the early nineteenth century and this could also have helped to shape the present-day Whitefaced Woodland. This breed has always been firmly associated with the hills between Manchester and Sheffield, the Hope Forest and High Peak area of Derbyshire, and local farmers kept the breed going without a breed society. By 1964 it was clearly in danger of extinction and it was one of the seven breeds of sheep included in the Gene Bank established by the Zoological Society of London at Whipsnade (Rowlands, 1964).

In 1971, after a lapse of twenty years, classes for Penistone sheep were included in the Penistone show. This was a significant point in the rehabilitation of the breed, which is now quite widespread, under the aegis of the Rare Breeds Survival Trust, and some flocks are of considerable size. Indeed, one registered flock numbers 270 ewes and is the largest flock of any single registered rare breed.

There is certainly a good deal of variation, but whether the Penistone is today a recognizably distinct type from the Whitefaced Woodland is debatable. It could be this confusion that underlies the coexistence of 1000 registered Whitefaced Woodland ewes and another 2000 unregistered.

Keith Huggett photographed these Whitefaced Woodland sheep (the Cherry Tree Flock of J.M. & V.J. Howard) at Holme, near Huddersfield, in 1985. The shepherd is James Howard.

Exmoor Horn and Devon Closewool

The hill sheep of south-west England in the eighteenth century were rather similar to those of Wales. They were usually horned, with white faces and legs. Their wool was not abundant and too hairy to be valuable. They owed some of their form to an unknown long-wool influence from earlier days and were at this point best described as a mixture of short-wools and middle-wools. The short-wools among them had died out in the western part of their range by the middle of the nineteenth century, while to the east they gave rise to a type known as the South-western Horn, the ancestral stock for today's Dorset Horn and Portland.

Devonshire sheep breeders of the late eighteenth century capitalized on the earlier work which had led to the middle-wool sheep. One variant was the polled Devonshire Nott, which by 1800 had been further developed by crossing with the New Leicester. This contributed in great measure to the Dartmoor, and thence to the Whitefaced or Widecombe Dartmoor and to

Shiels painted this picture of the Exmoor Horn, classified by Low as one of the 'forest breeds of England' – the ram having a beard, and the female 'like other wild breeds' being considerably smaller. He described the breed as climbing like goats and being very bold, 'often attacking sheep much larger than themselves'.

the Improved Dartmoor, also known as the Dartmoor Greyface or Dartmoor Longwool. Another route led to the Devon Longwool and the South Devon.

The other middle-wool variant was known to Charles Vancouver in 1808 as the Common Exmoor Sheep. It differed from the Devonshire Nott in having horns, and also became known as the Porlock, though originally this term was limited to the sheep from the north side of Exmoor. Not all commentators were enamoured with the breed. Coleman wrote in 1887 'Whatever may be thought of the Exmoor, sufficient merit is exemplified under cultivation to invest the sheep with high claims for perpetuity of existence'. When the Exmoor Horn breed society was formed in 1906, the higher quality and larger size of the southern variant was remarked upon. The present day breed, which does not seem to have changed very much from that illustrated in early flock books, represents the combination of these variants.

Exmoor Horns have customarily been taken off the hill for the winter and for lambing. They are not as hardy as the Scottish Blackface but continue to be numerous on Exmoor, where the

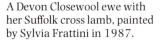

A Devon Closewool ewe with her Suffolk cross lamb, painted by Sylvia Frattini in 1987.

population is about 40 000. Attempts were made in the nineteenth century to introduce Cheviots to Exmoor but they did not achieve lasting success.

The long, soft wool is markedly less weather proof than the heavy fleece of the black-faced hill breeds, but the Exmoor Horn still qualifies for a hill sheep subsidy and this has helped to maintain numbers. Commercially, ewes are crossed with Suffolk rams for slaughter lambs, or with a Border Leicester for cross-bred ewe production.

The Devon Closewool arose around the middle of the nineteenth century when the Exmoor Horn was crossed with the Devon Longwool. Previously these sheep were known as Mongrels. The breed was developed as the western area of Exmoor appeared able to sustain a sheep intermediate in size between the Exmoor Horn and the Devon Longwool. A breed society was formed in 1923. The new breed, a grassland sheep, expanded very rapidly in numbers and by 1950 there were about 229 000, practically all in Devon. In contrast the Devonshire population of Devon Longwool and South Devon numbered 176 000; the Dartmoor 53 000; Exmoor Horn 30 000; Scottish Blackface 42 000, and cross-bred 206 000. The Devon Closewool looks very like a hornless Exmoor Horn.

The Devon breeds of sheep have not had much impact abroad. One of the very few records of exports after World War II is that of six female and one male Devon Closewool being sent to Canada in 1947.

10   Long-woolled breeds of sheep

During the earlier part of the eighteenth century in England generally there had been a shift from keeping unimproved short-woolled sheep on the poor and heath land, to long-woolled sheep on arable land, primarily for wool production. The sheep for this new husbandry were derived from the 'marsh' or 'pasture' sheep of Lincolnshire, Somerset, and Kent. Russell (1986) stressed that these little-documented changes in lowland sheep farming which took place before the 1780s were of the same degree of importance as the ones which occurred when the New Leicester and Southdown breeds spread over the country during the much better documented, later period of 1770 to 1850. With the industrial revolution, meat replaced wool as the important product of sheep, and the traditional long-woolled breeds declined.

There were originally five main breeds of these sheep; the Lincoln, Leicester, Teeswater, Romney Marsh, and Cotswold. These were all polled with white faces and legs, the wool was long and coarse and there was often a tuft of wool on the head which led to the name of 'mugs' or 'muds' for the Lincolns and Leicesters. They were very large sheep with a lot of bone and were slow to mature but they produced a fleece with wool that could be up to 15 inches (38.1 cm) in length. Trow-Smith (1959) and many earlier writers have suggested that the long-woolled sheep were introduced to Britain by the Romans.

By the middle of the eighteenth century the Lincoln and Leicester breeds were very similar in size, conformation, and fleece, and they were often interbred. It was from this stock that Robert Bakewell, by a strategy of close inbreeding and careful selection, produced his New Leicester or Dishley Leicester. The old Leicester breed became transformed before the end of the century into this new type, with a barrel-shaped body, a round broad back, and very short legs. This sheep fattened very quickly and was much smaller boned. It became a meat-producing animal rather than solely a wool-provider. Within a very short period the New Leicester had been crossed with all the other long-woolled breeds and with many other breeds of sheep in Britain. By the end of the century the old Leicester was nearly extinct, as was the Teeswater, and the Lincoln had become much closer to the New Leicester. The Cotswold and the Romney Marsh breeds, however, survived in much their old form, because the crossings with the New Leicester conferred no real advantage.

Leicester Longwool

The Leicester Longwool is descended from Bakewell's New Leicester (p. 151). The Colling brothers of Shorthorn fame bought rams from Bakewell, and their own ram sales were attended by Sir Tatton Sykes of Sledmere, who himself hired out rams until 1862. Mr H.P. Robinson of Carnaby, near Hull, began breeding Leicesters in 1795. These latter two flocks seem to have been the origin of the breed which, at a meeting held in Driffield, east Yorkshire in 1893, was given the name Improved Leicester. Twenty-three flocks and 104 rams were registered in the first Leicester Longwool flock book. The sizes of flocks ranged from 100 to 420 ewes and the dates when the foundation flocks were started went back to the 1790s.

The new breed was larger with a heavier fleece than the emerging Border Leicester and clearly filled a specialist role for farmers on the Yorkshire Wolds where hardiness and adaptability to arable farming were required. It seems highly likely that Lincoln crosses had contributed towards the new breed.

Judy Coleman with the champion Leicester Longwool ram at the Royal Show, York, 1948. The Speeton flock, where this ram was bred, is likely to date back at least to 1834 and continues to thrive.

Leicester Longwool ram, and a ewe of the coloured variety, displayed by Mr & Mrs J. Baylis of St. Albans, Hertfordshire.

The Improved Leicesters were first shown at the Royal Show at Darlington in 1895. The entry was not a success and the judge found it 'difficult to say what is the point of improvement aimed at'. Nevertheless the breeders were not discouraged and the Leicester Longwool has retained a place in East Riding arable sheep farming.

From early days the Leicester was advertised as being suitable for crossing with 'Colonial and Foreign Sheep' and an export trade to Australia and North and South America was built up. The adjective 'Improved' was removed from the Association's name in 1896 as foreigners were puzzled by it, and its use did not reflect the breed's long history.

In the 1920s and 1930s the Leicester did not prosper, as its carcass was too big and the wool market was very poor. However, the use of Leicester rams as terminal sires on Mashams and other cross-breds continued. The British sheep meat trade has traditionally been very complex with certain carcass types finding favour in certain areas. This was particularly true of northern England. Long, in his 1969 essay on Yorkshire farming, described how:

> the very small carcasses of fat lamb (28–32 lb [13–15 kg] dressed carcass weight) from Swaledale are particularly popular in Manchester . . . Leeds, however, is one of the few places which will take a 60 lb [27 kg] hogg from the arable districts in the spring . . . the industrial towns, both in the West Riding and on Tees-side, also provide a good market for ewe mutton.

The Rare Breeds Survival Trust surveys in 1974 and 1979 gave the numbers of Leicester Longwools as 235 and 327 ewes respectively and the breed has benefited from promotion and encouragement since then. A coloured variant exists in the breed, which was admitted to registration for the first time in 1986.

Lincoln Longwool

A large proportion of the English wool clip in the eighteenth century came from the long-woolled sheep of Lincolnshire whose fleece was 'wondrously heavy, and sends down long unctuous wool in pendulous masses almost to the ground'. On the poorer upland pastures, especially of the Lincolnshire Wolds, some 'little short-legged dwarfish sheep' were still kept into the middle of the eighteenth century but the main pattern of sheep husbandry was for long-woolled stock to be bred and raised on the higher ground and sold to graziers on the rich coastal marsh pastures.

Imports of fine Merino wool and the rapid development of the cotton industry, especially from the 1770s, caused a steady reduction in the profitability of British wool production, relieved by occasional good periods, as for example during the Napoleonic Wars.

In Lincolnshire by the 1780s the graziers faced particular difficulties because of their remoteness from markets and the competition from arable farms fattening sheep on clover, ley grasslands, and other 'new' crops. Furthermore, as Arthur Young wrote, 'The Lincoln has been entirely spoiled by breeding for quantity of wool only'. However, his pessimism proved unfounded as the breed was able to respond to the change in the market towards meat production. This improvement began in the 1760s and was said to be complete by the 1830s; the New Leicester as well as Merino rams were used. It seems quite likely that Lincoln rams had already been used by Bakewell himself in breeding his New Leicester, though he and many Lincoln breeders made a habit of denouncing each others' stock, sometimes in memorable terms. Charles Chaplin of Tathwell in 1788 described Bakewell's sheep as being 'without size,

'The Old Lincoln breed. Ram, bred by Mr Jex, St. Jermains, near Lynne, county of Norfolk.' In the 1840s several types of Lincoln sheep remained. On the northern Wolds the sheep were very similar to the New Leicester. In the central Wolds, flocks of a distinct, improved type were common. In the south the traditional sheep, bred for fattening on the marshes, was to be found. However, these differences were rapidly disappearing and the new breed was securing a place at home and abroad. This ram, painted by Shiels, bred by Mr Jex of St. Jermains, near King's Lynn, was closer to the old type of Lincoln than many others around when Low was writing in 1842.

without length and without wool'; Bakewell considered the Lincoln to have a 'barrowful of garbage' for mutton and he purchased an 'outstandingly ugly' Lincoln sheep to contrast with his own New Leicesters. At any rate, the new Lincoln was a heavy, hardy sheep which could spend the winter on turnip fields and consolidate the ground so that it was ready for cereal crops. Lincoln wool was suitable for the worsted and carpet making industries. Above all, the breed was an effective mutton sheep in terms of quantity of meat produced, albeit of low quality.

In sharp contrast to Bakewell and his associates, the Lincoln breeders did not use inbreeding to fix the type, and as a result the new breed was so variable in standard that it was only in 1870 that a separate class became established at the Royal Show. By this time an export trade had started, and by 1865 the Lincoln-Merino was a well known cross-bred in Germany. In the 1870s it was noted in Ireland, and the next decade saw its importation to New Zealand, where it was crossed with the Merino (imported 1850) to create the Corriedale. In 1892 the Lincoln Longwool Sheep Breeders' Association was founded.

By the 1970s the Lincoln Longwool was a vulnerable breed. Its big clip of wool, up to 15 kg of wool of 20 inches (51 cm) length on a ram, was not a protection from foreign competition, and other forms of farming had become more profitable. In 1907 there were estimated to be half a million Lincoln sheep in their native county, but by 1986 the breed numbered 829

Henry Dudding began farming in 1861 and was a most notable breeder of Lincoln Longwools from 1876 at Riby Grove. In 1906 his Royal Show champion was auctioned for 1450 guineas for export to Argentina.

ewes in 38 flocks. Pure-bred Lincoln flocks are certainly not the feature of the landscape that they once were, but about 42 tonnes of Lincoln and similar wool are clipped annually in Britain. This figure includes the clip from the related Teeswater and Wensleydale breeds, as well as from unregistered sheep of Lincoln Longwool type, and is small compared with that of, for example, the Romney Marsh. In 1986, eighteen times as much wool was clipped from the latter as from the Lincoln and related longwools.

Today there is still an overseas market for Lincoln rams; in 1986 160 rams were exported to Turkey, Bulgaria, and Poland, but this figure is in stark contrast to the 1906 export figure of 6928 breeding sheep.

Between 1790 and 1810, George Garrard made a remarkable series of plaster models of cattle, sheep and pigs, at a scale of two and a quarter inches to one foot. A collection of these survives in the British Museum (Natural History). These models probably represent shorn and unshorn Lincoln sheep.

Romney Marsh (Kent)

Romney Marsh occupies some 210 square kilometres of coastal Kent, but this relatively small area is the cradle of a breed that has exerted an enormous influence on the sheep industry of New Zealand, Australia, and South America. Between 1900 and 1955 over 18 000 rams and 9000 ewes were exported to 43 countries, but perhaps because the geography of the area has no parallel elsewhere in Britain, its sheep (while still one of Britain's most numerous lowland breeds with about 60 registered flocks and 7000 pedigree breeding ewes) have never had a marked influence away from the south coast. New flocks have been started in the West Country and Wales but the breed is still centred in Kent and Sussex. Unlike other long-woolled breeds, namely the Devon and Cornwall, Leicester, Lincoln, and Cotswold, the Romney Marsh has a secure place in British farming.

From about 1795, Richard Goord of Sittingbourne carried out much improvement by selection and inbreeding, and in the early part of the nineteenth century New Leicester blood was brought in, but a pronounced effect was not made upon the breed. The breed society was formed in 1895. A large body size (80 kg for a ewe) and heavy wool clip (3.5 to 4.5 kg, and up to 10 kg on a few individuals) are hallmarks of the breed which has a rather shorter fleece than other long-woolled breeds. Usually Romneys are crossed with a Suffolk, Southdown, or Texel ram for lamb production. Wool is an important source of income for the Romney flock and in 1986 about 800 tonnes of Romney wool was clipped in Britain.

The security of this breed in British farming is largely due to its identification with the traditional husbandry system of Romney Marsh, the only lowland part of Britain where sheep are the leading livestock. Defoe (1724–6) and other writers of the eighteenth century commented upon this husbandry system in which the different soil types and their varying drainage patterns and vegetation types were fully understood by the sheep farmers of the region, who adjusted the grazing patterns to suit individual swards. Abundant rye grass characterized the best pastures which were traditionally heavily grazed in summer, at a stocking rate of up to fifteen sheep per hectare. In winter, stocking rates were as high as the pasture could support, on average five sheep per hectare. Animals for which there was no grazing were maintained on farms elsewhere, no hay being provided in the pure application of the 'Romney Marsh system', which has been described as 'all-grass, all-sheep'.

Prolificacy was never emphasized by the breeders, who have always tended to select for hardiness, mothering ability, and longevity, and to accept a slow rate of maturing. A group breeding scheme is now operating and prolificacy is rapidly being increased. Indeed the traditional Romney combination of characteristics is reminiscent of hill breeds, and so too is the husbandry system of the Marsh. Like the uplands, the Marsh is a zone of production from where stock is taken for sale and fattening nearby. Also like hill-sheep, Romney Marsh sheep are often mated with crossing sires to produce grassland sheep, usually known as Kent Half-breds.

Little change seems to have taken place physically to the Kent or Romney Marsh breed between 1842, the date of Shiels' painting, and 1980, when R. Samaraweera painted this ewe (left). Archaeological evidence (the discovery of bones of very large, polled sheep of Romano-British, that is, of first or second century AD origin) suggests that the traditional idea of Roman ancestry could well be correct.

Cotswold

> The exact type of the original, unimproved Cotswold, with its great medieval reputation, is a matter of some doubt.

Thus wrote Trow-Smith in 1959. Certainly the Cotswold of the late eighteenth century had lost the fineness of wool for which it had been famed two hundred years previously. A major cause of this was the improved nutrition which lowland sheep received during the eighteenth century and, in places, in the previous century, as their husbandry became integrated with arable farming. The fleece of a well-fed sheep grows longer and has thicker fibres than those of a poorly fed sheep.

Before Bakewell had begun his improvements the Cotswold had been extensively crossed with the unimproved Leicesters. New Leicester influence bestowed good conformation, hastened maturity and restored some of the fleece quality. Breeders were anxious none the

Ralph Whitford painted Mr Garne's flock of Cotswolds near Northleach, Gloucestershire, in 1866. This flock was managed according to the traditional 'Golden Hoof' husbandry, this being heavy stocking of ewes and lambs, hurdled on fodder crops and roots, thereby enriching and consolidating the limestone soil for the arable part of the rotation.

less to maintain the breed's identity, and the historical preface to the Flock Book states that no Leicester blood was introduced after 1830. However, some flocks have cross-bred with Leicester Longwools in the twentieth century.

The Cotswold found valuable export markets in North America and Australia, where these big, long-legged, heavy-fleeced rams supplied the same kind of market as the Lincoln and Romney Marsh. Back home, mutton quality was too poor to compete initially with that of the Down breeds and latterly, with frozen imported meat. Breeders of the Romney Marsh had faced up to the same problem by adopting the Southdown as the terminal sire, retaining its identity as a pure-bred, but this did not happen with the Cotswold where the Southdown and Hampshire Down crosses were so vigorously applied that an altogether new breed, the Oxford Down, emerged. By 1949 there was only one flock of Cotswolds, owned by the Garne family. In an interesting twist to the tale of Down breeds supplanting the Cotswold, one of the best markets the Garnes had in the 1930s and 1940s for their rams was for use on Suffolk ewes in East Anglia.

Today the Cotswold is increasing slowly in numbers as farmers are finding that the slightly higher cost of keeping these sheep, as opposed to cross-breds, is compensated by the pleasure of helping to conserve the old breed.

Devon Longwool and South Devon

The Devon and Cornwall Longwool Association was formed in 1977 when the breed societies for the Devon Longwool and the South Devon decided to merge. The two breeds had a common origin; in the eighteenth century local races called the Bampton Nott and the Southam Nott had emerged, presumably as a result of crosses by longwools from elsewhere upon the native sheep. Further development involved the use of New Leicester and probably Lincoln rams, in order to produce remunerative sheep adapted to the new turnip husbandry of south Devon as quickly as possible. The local improvers were not discouraged by the earlier failures of the New Leicesters provided by Bakewell and his followers whose unhappy experiences were reported by Robert Fraser, surveyor for the Board of Agriculture, in 1794.

The minute book of the Netherexe Farmers' Club and Agricultural Discussion Society survives to tell us that, in 1839, 'it is the opinion of the Society that Sheep should feed off common Turnips until the latter end of January except the weather be very severe. Also that the Leicester Breed of Sheep is the best for this neighbourhood'. By 'this neighbourhood' was meant the country between Exeter and Tiverton.

The breed became differentiated into two races; the South Devon, occupying the east and south of Cornwall and south Devon which was rather larger and longer in the leg from selection having favoured a heavy wool clip, and the Devon Longwool, found in north Devon, north Cornwall, and south Somerset. Both were arable sheep and both lost ground steadily with the increase in dairying from 1875.

In many ways the breed today resembles the Lincoln Longwool, having a big-framed body and a massive clip of wool. Many flocks average 8 or 9 kg per fleece, with 12 inches (31 cm) of

This ram, Glory, painted in 1854, was the property of Thomas Potter of Thorverton, Devon, who exhibited long-woolled sheep at the Bath & West Society Shows during the 1850s. At this time such sheep could be entered in one of several classes: 'Leicester', 'Cotswold', or 'Long-woolled not qualified to compete as Leicester or Cotswold'. Glory was probably the ram with which Potter won second prize in the last-mentioned class in the 1854 Show, and was a forerunner of the distinct Devon Longwool breed.

Devon and Cornwall Longwools in their native country, near North Tawton.

staple. Individual rams have clipped up to 18kg, and ewes up to 13.5 kg. Most of the wool is used in carpets. 'Southwest Cross-bred' is a distinct description of wool most of which comes from the progeny of a Devon and Cornwall ewe and a Down ram, traditionally the Dorset Down. The annual clip of about 1100 tonnes of this wool is half as much again as that of the Romney Marsh. Unusually for a British breed, Devon and Cornwall Longwools are shorn as lambs, producing a particularly warm, resilient, and valuable wool.

Dartmoor breeds

The Dartmoor Sheep Breeders Association was founded in 1909 and the breed it cares for is known as the Improved Dartmoor, Dartmoor Greyface, or Dartmoor Longwool. The type kept around Widecombe became known as the Whitefaced Dartmoor and a breed society was formed for it in 1950. The Whitefaced Dartmoor is accepted to be closer to the primitive native sheep of the Devon hills. Rams are sometimes horned, and the breed is more 'moorland' in character. The Dartmoor Greyface is a polled breed and this itself suggests a more extensive influence of improving breeds. Youatt (1837) and other historians of the period, summarized by Thompson (1984), cite the Leicester, Cheviot, Southdown, and Lincoln as influences. The speckling of the face and legs in particular suggests the Southdown influence.

Within Devon there are no fewer than five native breeds of sheep, each with a different history and tradition. Perhaps this is not altogether surprising in a county with so many small and isolated farms, but this diversity should certainly not be taken for granted, and it is important that all these local breeds continue to thrive. There is stiff competition from other breeds of sheep, and in one sense the Dartmoor breeds are misnamed as the only breed of sheep existing in any numbers on Dartmoor is the smaller and hardier Scottish Blackface, which has found favour there since the 1930s because it can be stocked at higher density in summer and can overwinter on the moor. Changes in arable farming have meant there is now less winter fodder available for sheep, and the Dartmoor breeds depended on this.

Since 1945, the expansion of dairying and the ineligibility of the Whitefaced Dartmoor and Dartmoor Greyface for a hill ewe subsidy have acted against their popularity. Today neither is strong in numbers, particularly the latter which in 1986 had about 23 breeders, 852 registered females and 260 ewe lambs registered annually. Including non-registered stock the total is probably about 2000 sheep, and the breed is now on the Priority List of the Rare Breeds Survival Trust.

Old livestock photographs often serve to show changes in breeds. These pictures (dated 1923 and 1953) show how the Dartmoor Greyface has become slightly stockier, and in recent years this trend has continued.

11   The Merino experiment

> The preservation of the Merino race in its utmost purity at the Cape of Good Hope, in the marshes of Holland, and under the rigorous climate of Sweden furnish an additional support of this, my unalterable principle, fine-wooled [sic] sheep may be kept wherever industrious men and intelligent breeders exist.

'The Merino breed. Ram & ewe, bred by Mr Benett of Pyt House, Wiltshire, M.P.'
The Merino failed to become established in Britain, and cross-breds with a high proportion of Merino blood were not successful either. Even so, Trow-Smith (1959) stated that many, if not most British sheep breeds received some Merino influence. This was most successful with the North Country Cheviot.

This pronouncement, by M. Lasteyrie (an 'unwearied advocate' of the Merino) was reported by Youatt in 1837. Today these words would be acclaimed in Australia where the Merino provides the greatest proportion of the world's supply of wool, but the breed has never been very successful in Britain despite its early, royal patronage.

Merino sheep came originally from Spain where their fine wool had provided an industry on which the fortunes of the country depended since the middle ages, for it was in great demand throughout the rest of Europe. Until the end of the eighteenth century the Spanish guild, the Mesta, forbade the export of Merino sheep but with the beginning of the industrial revolution, the development of looms and of techniques of weaving meant that there was an ever increasing demand for the finest wool. The King of Spain was one of the first to allow the expansion into foreign countries. Merinos reached Germany in 1765, France from 1767–86, Britain in 1785–7 and South Africa in 1789, from where they were taken to Australia.

The first sheep to reach Britain were a pair obtained from France by Sir Joseph Banks in 1785. Then in 1789 George III was persuaded to obtain some and a small number were more or less smuggled into England from Spain across the Portugese border. These were a motley collection and did not impress the King, who then approached the King of Spain directly for specimens from his prize 'Negretti' type of Merinos. These quickly supplanted the earlier imports and formed the basis of the British stock which was patronized by Banks, Coke of Norfolk, and Lord Somerville, who attempted to cross-breed their Merinos with British breeds. A Merino Society was founded in 1811 but in the same year Coke admitted to a meeting at Holkham that the crosses born from his Southdown ewes that had been put with Merino rams were not thriving. Crosses with English longwools did not produce the desired early-maturing, hardy and fine-woolled sheep, which was surprising in view of the fact that today the Merino-Lincoln cross, which became the Corriedale, is the most important breed on South Island, New Zealand, and on the Falkland Islands.

The white, painted face, fine long horns, and heavy neck suggest this model represents a Merino and it may be the same ram as is illustrated in Garrard's well-known painting of 1804, the *Woburn Sheepshearing*.

Despite its failure in Britain the Spanish sheep elsewhere has, either directly or indirectly, effected a complete revolution in the character of the fleece, which Youatt would be gratified to see. He would also be pleased to note that interest in the breed persists among British sheep farmers; the remarkable prolificacy of the Booroola Merino is being studied experimentally and Merinos are kept on a small scale in Britain for their wool.

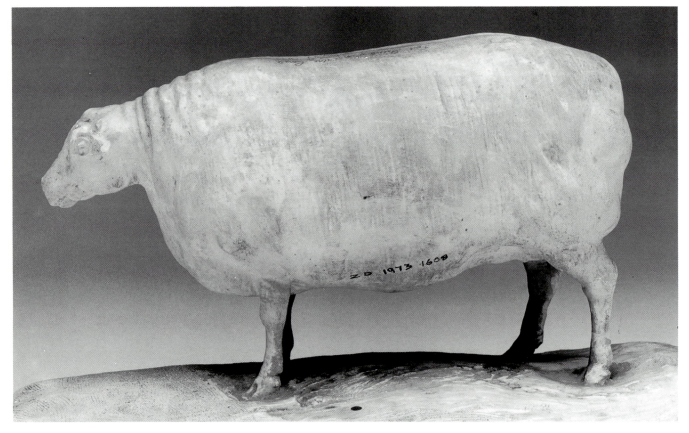

12 New Leicester and the crossing sires

'The New Leicester breed. Ram, a two-shear sheep, bred by Mr Buckley, Normanton Hall, Leicestershire.'
These sheep were criticized for their heavy forequarters. Bakewell countered this by arguing that the cheap cuts on these parts appealed to the poor who, rather than the rich, were customers for fat mutton. This picture shows the early stages of the development of the Dishley Leicester into the Border Leicester.

The New Leicester sheep can be taken as the symbol of the eighteenth century livestock improvers because it was on this breed above all others that Robert Bakewell established his principles of inbreeding to produce a new type of livestock animal. Beginning about 1760, over a period of twenty to thirty years, Bakewell changed the Leicester breed from a large-boned, long-legged sheep into a light-boned, barrel-shaped animal with short legs. It was claimed that the New Leicester matured about a year earlier than the old breed and it soon became renowned as the perfect butcher's sheep.

Having established his flock of New Leicesters, Bakewell began the practice of letting out rams for the season, although at first he had great difficulty in persuading farmers to accept this new and innovative idea. The first ram was let in about 1760 for sixteen shillings; by 1780 he was receiving ten guineas for each ram. In 1789 Bakewell made 1200 guineas from three rams and 2000 guineas from seven others, but his most remarkable letting was of his favourite, Two-pounder, at 1200 guineas for one season. The ram's name was due to his resemblance in shape to the barrel of a two-pounder cannon.

In 1837 Youatt wrote the following account of the society that was formed by Bakewell and others for the purpose of ram-letting:

> After the reputation of Mr Bakewell's flock had been established, and its superiority over that of any other breed then existing very generally admitted, he was enabled to establish the Dishley Society. There can be little doubt that the object which he had principally in view in forming this Society was to promote his own interest, and that of the members of the Society, and its establishment has very mainly contributed to preserve the purity of the new Leicester breed, and has thus been very beneficial to the country. Its principal rules are, –
>
> 1st. No member shall hire or use a ram not belonging either to Mr Bakewell or to one of the members of the Society.
>
> 2nd. No member shall give his rams, at any season of the year, any other food than green vegetables, hay, and straw.
>
> 3rd. No member shall let more than thirty rams in one season.
>
> 4th. No member shall let a ram for less than ten guineas to any person, nor for less than forty guineas to any person who lets rams . . .
>
> 11th. From the 1st to the 8th of June the members shall not show their rams, except to one another. They shall begin their general show on the 8th June, and continue to show their rams till the 8th of July.

This model bears the label, 'Fat Leicestershire Ewe, G. Garrard, May 31st, 1810'.

By the end of the century the New Leicester had been interbred with almost all breeds of sheep in England but it never replaced them, and the tendency of the breed to become excessively fat acted against its continuing popularity. Today the most notable descendants of Bakewell's stock are the Border Leicester and the Leicester Longwool.

The Border Leicester and certain other breeds which also owe a good deal to Bakewell's endeavours, play a vital role today as crossing sires. Their daughters, produced when they are mated with hill ewes, are the cross-bred ewes which combine the dam's hardiness with rapid growth and fecundity conferred by the sire. Crossbred ewes are Britain's main lowland sheep and they are mated with meat sires, also known as terminal sires, to produce meat-lambs for slaughter.

Sir John Palmer of Withcote Hall, Leicestershire inspects his New Leicester sheep, with shepherd John Green, painted by John Ferneley in 1823.

Border Leicester

The oldest Border Leicester flock in Britain, at Rock near Alnwick in Northumberland, was founded in 1849. Until the 1920s flock policy was to buy in no females and very few males (from 1856 to 1896 only sixteen rams in total were bought in). At its peak the flock numbered 400 sheep. Sixty rams were sold each September at Kelso and the same number at other northern sales.

In the 125 years up to 1972 when the history of the flock was written, there had been a total of only six head shepherds at Rock. Some of these shepherds are illustrated here and on p. 155.

Right: the three-shear ram, Sir Walter, shown at the Highland and Agricultural Society's Show at Edinburgh in 1869 by the Revd R.W. Bosanquet. It gained first prize in the class not above four shear. With the ram is Mr James Aitchison, agent for Rock from 1846 to 1877, and Andrew Thompson, the shepherd at that time.

Two of Bakewell's followers and friends in the early 1760s were Matthew and George Culley, from County Durham, who moved to Fenton in north Northumberland in 1767. They developed a famous flock of New Leicesters by crossing with the Old Teeswater, a long-woolled breed which was soon to die out. In 1792 they were involved in the foundation of the Milfield Society, the local association of breeders of New Leicesters, and which before long was in competition with the corresponding Leicestershire-based society over the hire of rams to Yorkshire breeders. Indeed, after Bakewell's death in 1795 the Culleys had very little to do with the southern breeders. Other northern breeders maintained links, but not after about 1830.

This isolation, together with the influence of the infusion of blood from the Old Teeswater (described by George Culley as 'confessedly the largest sheep in this Island, but far from handsome') helped the New Leicester quickly to establish its distinct identity. So far as is known, blood from this new breed did not influence the purebreeding local Cheviots appreciably and any part played by the latter in the Leicester's development, though likely, remains undemonstrated. McDonald, writing in 1909, was of the opinion that numbers of flocks and

The Bluefaced Leicester arose in the late nineteenth century, although the breed society was not formed until 1963. Pawson (1961) describes how at Hexham market it was possible at the same sale to select the dark-faced Hexham Border Leicester or the white-faced Border Leicester proper (known there as the Kelso type). Pawson repeats the suggestion that the dark-faced type, later to become known as the Bluefaced Leicester, fetched lower prices and therefore tended to be bought by the less wealthy hill farmers who kept Blackfaces or Swaledales; selection therefore proceeded along the lines of a crossing sire for these breeds. Today this cross is called the Mule, and is now the premier cross-bred in Britain. The main sale centre is now Lazonby, near Penrith, Cumbria, where up to 26 000 breeding Mule lambs change hands on each of the sale days in September and October. Painted by Alan Stones in 1984.

Right: Rock Reward, a Border Leicester, was sold at Kelso in 1948 for £1200. The shepherd was Walter Walkers.

Below: Robert Duncan, shepherd at Rock from 1893 to 1917, with the Border Leicester ram, Rock Royal, in 1902.

Below right: Mr A. Watson (manager) and Archie Plenderleith (shepherd), Rock, 1969, with a Border Leicester ram.

155

of individuals of the new breed grew at such a rate that local ewes must have been used in grading-up programmes.

Although George Culley had acquired a very high reputation as a sheep expert by the time he died, aged 78, in 1813, local sheep farmers, committed to Cheviot and Blackface sheep, were not in favour of the new breed. One landowner, William Mure, reported by Rowe (1971), expected to put a ram hired from the Culleys to his ewes, but 'the inveterate enemies to *Cullies sheep* turned in a blackguard Galloway ram amongst the flock at a distant farm on hand, out of the whole flock only five could be found that had not got the ram . . .'

The new breed was recognized by the Highland and Agricultural Society in 1869, and it became known as the Border Leicester. Some time previously it had been found that commercially this breed and the hill sheep had a great deal to offer each other and there was no need for one to replace the other. This was because when the Border Leicester is crossed with the Cheviot, the result is a valuable cross-bred known as the Scotch Halfbred. The cross-bred with the Blackface became known as the Greyface. Certainly since 1850 the sale of cross-bred sheep, initially for the butcher in Edinburgh, but later for lowland sheep farmers to use as breeding stock, has been very important. Wherever the Blackface and Cheviot went, the Border Leicester was not far behind. Pedigree flocks of the new breed were kept under careful husbandry conditions to provide rams for crossing with the Blackface and Cheviot flocks; the first flock book was published in 1898. The main function of the Border Leicester is to pass on the trait of large litter size and rapid growth to the cross-bred progeny.

The main concentration of mountain sheep is in Wales, and perhaps it is surprising that it was only in 1907 that the first registered Border Leicester flocks were founded there; the Welsh Halfbred (progeny of the Border Leicester and the Welsh Mountain ewe) is now an important lowland sheep. The breed was first exported to Australia in 1871 and is used for crossing with the Merino. The 'First Cross' ewe resulting from this mating yields 75 per cent of the 16 million prime lambs slaughtered each year in Australia.

Wensleydale and Teeswater

TEES-WATER OLD OR UNIMPROVED BREED.

TEES-WATER IMPROVED BREED.

From *A general history of quadrupeds*, by Thomas Bewick (1790).

Youatt (1837) considered that the polled, white-faced Teeswater sheep, which had been restricted to a confined district along the river Tees on the Yorkshire—Durham border, were of the same stock as the old Lincolnshires. Local selection during the late eighteenth century and an admixture of New Leicester blood improved the breed, which had already been kept for some time under the good conditions of generous feeding which enabled its fecundity to be developed. Culley recorded in his book (1807) how one Teeswater ewe had dropped twenty lambs in six years. The breed retained high fecundity and with its lustrous long-woolled fleece gave rise to the Wensleydale breed which is characterized by a deep blue head and ears. This feature of the breed is ascribed to the famous ram Bluecap, a blue-headed ram born in 1838 to a Teeswater ewe belonging to Mr Outhwaite of Appleton who had hired a large Leicester ram from Mr Souly, a well-known breeder of Leicesters. Today, the Wensleydale still has the distinctive fine, lustrous fleece, and blue head of the founder animal, Bluecap; few, if any other breeds of sheep owe their origin to a single sire.

By the time Bluecap came on the scene the Teeswater as reared in Wensleydale was known widely as a successful crossing sire for Swaledale type sheep. The newly acquired colour-marking propensity was a marketable feature as blue headed rams sired lambs with a darker mottled face and legs, which commanded a higher price.

In 1876 the Yorkshire Agricultural Society began to give prizes for the breed which acquired the name Wensleydale at that time. The first special classes were held at the Royal Show in 1883. Two breed societies were formed in 1890, which merged in 1919 after differences of opinion over acceptance of foundation rams for registration had been resolved.

TEESWATER

SINGLE PURPOSE
DUAL RESULT
RAMS
bred for results
—judged by results

The ideal breed of
Ram for Crossing

TEESWATER AGED RAM
IN FULL WOOL

Sires of

TEESWATER
HALF - BRED

MASHAM

MORE PROLIFIC
MORE PROFITABLE
EWES
have bred and reared
the highest lamb crop
in the National Lamb-
ing Competition
every year since 1956

MASHAM GIMMER HOGGS
IN FULL WOOL

More Lambs per Ewe give More Profit per Acre

Teeswater promotional
literature – 1962.

Very soon the Wensleydale was being bred almost entirely to provide crossing rams and in the 1920s most of the registered flocks comprised fewer than twenty ewes. The ability of the breed to confer good quality of fleece became well known and attempts have been made (notably in Africa) to upgrade hairy-coated sheep into wool breeds by crossing with Wensleydales.

Rams of the new Wensleydale breed sired, out of Swaledale or Dalesbred ewes, the Masham cross-bred. Masham ewes have been an important lowland sheep for over a century particularly in the western parts of Yorkshire. The Wensleydale retains its ability to confer fleece quality on its progeny, and in spite of the hairy fleece of its Swaledale or Dalesbred dam, the Masham has little or no hair in the fleece. The cross of Masham ewe with Wensleydale ram gave the 'twice-crossed' lamb, which yielded in popularity to the Oxford Down-sired cross. In the late 1950s this ram breed was superseded by the Suffolk, but it has always been a feature of sheep farming in Yorkshire that different markets favour different kinds of lamb so a considerable complexity of cross-breeding systems remains.

The Wensleydale probably also contributed to the Bluefaced Leicester, which is mainly derived from the Border Leicester. When crossed with black-faced hill ewes, progeny known as Mules are produced, the most important cross-bred sheep of the present day in Britain. The North of England Mule is the progeny of the Swaledale ewe while the Highland Mule has the Scottish Blackface as its dam.

In 1922 Henry Robinson commented on the establishment of some all-black Wensleydale flocks for ornamental purposes. He stated that black Wensleydales were, indeed, often preferred by commercial farmers. Genetic studies show the black colour to be controlled by a recessive gene, presumably associated with the blue face. It was only in 1983 that the breed society agreed to register coloured Wensleydales, but to prevent the coloured gene entering white flocks the white offspring of coloured sheep are not accepted.

Wensleydale ewe and pure-bred lamb. Shorn of her lustrous fleece the Wensleydale ewe is a long-necked, long-legged and heavily jowled animal; definitely not a butcher's sheep. The main role of the Wensleydale has been its contribution of mothering ability and fleece quality to the famous Masham cross-bred sheep.

Competition from the Bluefaced Leicester and the Teeswater was heavy, and by 1977 the Wensleydale Longwool Sheep Breeders Association was not thriving, with only some 300 ewes. Since then the breed's fortunes have revived and there are now nearly 1000 ewes and over 60 flocks. While the market for crossing rams is probably saturated now, the Wensleydale represents a most valuable genetic resource on account of its fleece characteristics. It has the finest of the longwool fleeces, the fibres growing very rapidly and bearing a lustre, or shine, due to the smoothness of the surface scales on each fibre. Its wool is of high quality and is often mixed with mohair to make knitting yarns.

It seems that a few original Teeswater sheep survived in the further reaches of the Dales without receiving Bluecap's influence and in the late 1940s these remnants were consolidated into the Teeswater breed, characterized by the ancestral feature of white face and ears with black spots. The Teeswater flock book, established in 1949, contained 185 flocks. Unlike the Wensleydale, the Teeswater is today not a rare breed. This is because Mashams tend to be sired by Teeswater rams rather than by Wensleydales, apparently because the progeny have a more mottled face, which is preferred by sheep farmers. In his 1969 report on Yorkshire farming Long described the two breeds as 'today distinct, though commercially their similarity to each other is something like that of two peas'.

Three Masham ewes; watercolour by Shelagh Powell (1987).

13   The rising stars

British sheep breeding in recent years has mainly been directed towards increasing prolificacy of the ewe, and ensuring that the carcass of the lamb has a minimum of fat and bone. Within a given breed, the number of lambs dropped at one birth is of low heritability, that is the differences which may exist among ewes are not genetic, but are mainly a consequence of environment and management. This means that within-breed selection for prolificacy is not likely to be very effective, as earlier generations of sheep breeders have already raised the fertility of ewes as high as is practicably possible for each individual breed.

The fecundity of the cross-bred ewe is more amenable to improvement. This is because the genetic factors involved can act additively and it is perfectly reasonable to expect to be able to add prolificacy from one breed to, say, hardiness or meat quality from another. For the breeding of cross-bred sheep, the types of ewes available are fixed, being mostly old (draft) hill ewes, and cannot in practical terms be changed much (although individual Scottish farms might for instance introduce Swaledale blood to their Blackface flocks, or change from Cheviot to North Country Cheviot). There is, however, a very wide range of possible crossing sires available and the financial rewards for breeders of a new crossing sire are potentially great. Likewise a new terminal sire which produces rapid-growing lean lambs could be very remunerative.

The British Friesland (East Friesian), British Milksheep, Bleu du Maine, and Rouge de l'Ouest are some breeds introduced and developed in the last thirty years to try to take over as crossing sires. Other imports, notably the Texel, Charollais, Charmoise, Ile de France, Vendeen, and Oldenburg were brought in principally to serve as terminal sires.

The validity of the claims made for these breeds is very variable, and a great deal of the evidence for or against particular breeds is anecdotal and not available for scrutiny. However, the value of the Texel as a sire for lean meat is indubitable although earlier claims for it as a crossing sire have not been substantiated. The success or failure of a breed may depend critically on the quality of husbandry, as is the case with the East Friesian which is prolific with high milk yield and a heavy body weight, but it is rather delicate and intensive management is needed to exploit its advantages.

Most, if not all of the imports named here have some ancestry attributable to British long-woolled breeds. The Texel arose from indigenous Dutch sheep crossed with Lincolns and various Down breeds. Interestingly, several of these new breeds are now being promoted for sheep dairying, although many of the 5000 or so milking ewes in Britain (in 150 flocks – 1985 figures) are of more traditional types such as Clun Forest and Mule.

Some of the new breeds have, in contrast, been entirely developed in Britain. In the late 1960s and early 1970s the Cambridge emerged. Selection was at first for reproductive performance, latterly for improved growth rate and milk yield. Starting in 1964, ewes were collected which had produced three consecutive sets of triplets. Most of these were of the Clun Forest breed, but there were some Lleyn, Llanwenog, and Kerry Hill. Hill Radnor, Ryeland, Border Leicester, and Bluefaced Leicester also had some influence. Seven unrelated Finnish Landrace rams, each mated with six ewes, were used for the first cross; subsequently back-crossing to the original British breeds was carried out and now the breed is about three-quarters British and one-quarter Finnish.

Advertisement from British Texel Sheep Society brochure, 1986.

The value of the Finnish Landrace (Finn) as a prolific sheep first came to general notice in Britain in 1962 and sheep were imported by the Animal Breeding Research Organisation in Edinburgh. Commercially the pure breed was not well accepted because it gave too many lambs, with higher mortality and low growth rates. Lambs fetched poor prices because they were of a new and unfamiliar kind even though carcass weight, conformation, and grade were satisfactory. The Finn Dorset, a cross with the Dorset Horn, has, however, secured a place in intensive sheep farming.

The Animal Breeding Research Organisation integrated the Finn's prolificacy with characteristics of other breeds to produce the Damline, which is, however, not recognized as a breed. The line was closed in 1973 and, genetically, the final composition was at that time: Finn 47 per cent, East Friesian 24 per cent, Border Leicester 17 per cent and Dorset Horn 12 per cent. The East Friesian conferred high milk production, litter size and body size, the Border Leicester also contributed the latter two characteristics, while the Dorset Horn was known to cross well with the Finn, as well as to extend the breeding season.

In contrast to the Damline the British Milksheep and Colbred have been enthusiastically promoted and marketed, leading to new breeds which are intended to increase the prolificacy and milk yield of the lowland cross-bred flock. They are also finding favour as dairy sheep. The former involved the crossing, by Mr Lawrence Alderson, of five breeds, including the Lleyn. The Colbred was developed by crossing Border Leicester, Dorset Horn, Clun Forest and Friesland, beginning in 1956 with the importation of Friesland sheep. In the words of its creator, Mr Oscar Colburn of Crickley Barrow, Gloucestershire, 'Each was crossed with each other and the best of the first generation were then mated together, so that each second generation contained 25 per cent of the original breeds. These hybrids were then inbred for ten generations to create a homogeneous population.' (Colburn pers. com.)

14 Upland sheep

In this category has been placed a number of hill breeds from England and Wales which have distinctive characteristics and histories. The Portland, now a very rare breed, is a relic of the old tan-faced race of primitive sheep that was common throughout Britain, but mainly centred in this form in the south-west. Sheep of the Portland type were probably ancestral to the Dorset Horn, a breed that has been renowned for at least 150 years for its early lambing and extended breeding season. The old breed of Dorset horned sheep was said to be superior to that of Wiltshire, being more compact and with shorter legs. Both were white sheep with large twisted horns, but the Old Wiltshire Horn was tall, long-legged, big-boned, and narrow-nosed. Today it is unique among British breeds in the nature of its fleece which is self-shedding.

The Dorset Horn and the Wiltshire Horn are both breeds from the chalk Downs of the south; to the north-west there is another group of upland sheep also with distinctive characteristics. The Ryeland breed originated in Herefordshire and during the eighteenth century it was the only sheep that was reputed to have wool which approached that of the Merino in quality, although at this time its numbers greatly decreased owing to the rising popularity of the Cotswold. Sir Joseph Banks, the promoter of the Merino in Britain, believed that the Ryeland (or Archenfield as he called it) when crossed with the Merino would produce the finest wool in Britain and he attempted to foster this cross but it appears to have found little favour.

Other breeds of the area are the Clun Forest, Hill Radnor and Kerry Hill. These are also breeds of the Welsh borders and the Kerry Hill is said to have evolved in the area of the same name in Montgomeryshire. The history of these breeds is as Trow-Smith (1959) described it, 'a tangle of horns and polls; black faces, white faces, and intermediate types'. The present day improved breeds bear little resemblance to their nineteenth century progenitors and the same is true for the Shropshire which was derived from a mixture of these old upland sheep and the Southdown.

Created around 1800, probably in Somerset, by an unknown artist from the coloured sands of the Dorset coast, this remarkable picture depicts horned sheep with a dished face, similar to today's Portland.

Dorset Horn and Portland

The Dorset Horn is a unique breed in several ways. It is the only British sheep which has the remarkable ability to lamb out of season. This has led to its reputation for producing two crops of lambs in one year. It is unusual, historically, for a lowland sheep in not being visibly influenced by the New Leicester or the Southdown. In recent years its combination with the Finnish Landrace (notably prolific, but breeding late in the year) has resulted in a remarkable cross-bred, combining the virtues of the two breeds. This is the Finn Dorset, which when subjected to modern reproductive technology can give 3.3 weaned lambs per ewe per year.

Lisle's *Husbandry* of 1757 describes how in Essex at that time 'Dorset house lambs' were produced in late autumn (from a March—May conception), housed, and fed by foster mothers in the daytime, the natural dams being brought in to suckle them at night. They supplied the rich London Christmas lamb market until the middle of the nineteenth century.

The Dorset Horn has strong historical affinities with the other horned, white-faced short-wools of the West Country, the Wiltshire Horn and Exmoor Horn, and is not far removed from the Welsh Mountain. The Portland is probably a relic of the common ancestral stock. However, none of these breeds shares the reputation of the Dorset Horn for out of season lambing.

More correctly referred to as an extended breeding season, the Dorset Horn's receptivity to the ram begins in July whereas under the same conditions other breeds come into heat in October. This was clarified in trials at Cambridge in the late 1940s where groups of 20 to 30

The property of Mr Michael Miller, of Plush, between Dorchester and Sherborne; these sheep, painted by Shiels, were described by Low as the 'last pure flock of original Dorset Horn sheep in the kingdom'.

166

ewes of mixed breeds ran together with vasectomised rams, on pasture, all year round. As with all other breeds, the Dorset Horn comes out of breeding condition in February or March. A few ewes may come on heat in April or early May in which case they are experiencing the end of their breeding season, and it is not possible to prolong it as a regular practice. Those coming on heat in late June are at the commencement of theirs. Matings earlier than July have to be contrived. This is done by exploiting the 'ram effect'; that is, the sudden introduction of a ram to ewes leads to otherwise silent heats being expressed, and conception follows. For reliable results, however, the ewes must be primed by hormone sponges, and one ram must be allocated to ten ewes, as opposed to one ram to around forty ewes in conventional autumn mating.

John Young's Ram Brewery, Wandsworth, has a Dorset Horn ram as a company mascot. The brewery was built on the Ram Field where the villagers of Wandsworth may at one time have kept a communal ram.

The extension of the breeding season in the Dorset Horn has had the extra time added to the start of the breeding season rather than to the end. This is characteristic of sheep breeds which originated in less northerly latitudes (30 to 40 deg. N) including the Merino, and while there is no direct historical evidence that the Merino contributed to the Dorset Horn, this distinctiveness of breeding season argues very strongly for such a connection.

Attempts were made in the early nineteenth century to add New Leicester blood to the Dorset Horn but they were unsuccessful. Similarly the Southdown made no impression on the breed. In the last thirty years, however, the breed has been transformed, outwardly at least, by infusion of Poll Dorset blood from Australia. The Poll Dorset was developed in Australia, starting in 1937 when a Corriedale ram was accidentally mated with a Dorset Horn ewe giving a hornless ewe lamb. This gave rise to the Newbold Polled Dorset flock. In 1941 the Valma flock was started, with Ryeland ewes being mated with a Dorset Horn ram. The process was expensive, difficult and controversial, but by 1954 the desired result, a polled breed fully comparable in carcass quality with the original Dorset Horn, had been achieved and the breed was firmly established.

The Portland breed was, in the early nineteenth century, only to be found in large numbers on the Isle of Portland in the south-west of the county. Sir George Crewe described in his journal a visit paid to Portland in 1835 to buy additions for the flock established by his grandfather at Calke Abbey, Derbyshire. In his eyes the Portland sheep were 'Dorset Horned sheep . . . starv'd into a smaller Compass'. The present-day breed is descended by a variety of routes from the Calke Abbey or Harpur-Crewe stock, and is one of Britain's rarest sheep.

Young Reggie Benham, of the Bath family of butchers, with a fine ram of distinct Dorset Horn type. In addition to its value as a meat sheep the Dorset Horn produces particularly fine wool which grows very densely, so the breed is well suited for the production of sheepskins.

Conservation has not been easy; the ewes seldom produce twins and there is no obvious commercial potential for the breed or for its cross-breds. The breed shows no tendency to have an extended breeding season, and may represent a pre-Merino Dorset Horn. In 1972 all Portland flocks (there were only about six) were traced on behalf of the Rare Breeds Survival Trust, the owners persuaded to register the sheep, and breeding plans proposed. The aim was to conserve genetic diversity by encouraging outbreeding among flocks and to ensure that as many as possible of the 35 rams and 107 ewes in the foundation stock which were judged fit to breed from continued to be represented in current lamb crops. Indeed, when pedigrees of all the lamb crops from 1981 to 1984 were drawn, around 40 per cent of the foundation males and 35 per cent of the foundation females were still represented.

The recent history of the Portland shows how genetic conservation can be achieved without the imposition of a mating plan, which genetical theory would require to involve random selection. The prevailing regime is one where animals are in private ownership, with advice provided by the Rare Breeds Survival Trust as a central organization for breeders.

A Portland ewe and her lamb. The fox-red colour of the latter is distinctive.

Kerry Hill, Hill Radnor, Clun Forest, and Llanwenog

Davies, the Board of Agriculture's surveyor of North Wales described in 1810 the breed of the Kerry Hills as being 'perhaps the only species in North Wales which produces perfect wool'. It was also 'comparatively tame and not so disposed to ramble' as other hill sheep. While it could have been a direct forerunner of today's Kerry Hill, other local races could have contributed as well. Thomas Halford, the breed's historian, wrote in 1898 that by 1855 the type was fixed (probably with the addition of some Shropshire blood) and rams were being sold to sheep farmers in the Clun Forest area. The flock book was established in 1899 with twenty-three registered flocks.

The Kerry Hill has been exported quite widely to South Africa, Europe, and elsewhere, but it has not had a very great influence abroad compared to that of, for example, the Shropshire. In Britain the Kerry Hill has been involved in a remarkable range of different husbandry systems. In its native area before 1939 it acquired a function as a crossing sire. Welsh Mountain ewes were crossed with Kerry Hill rams and the produce sold to other farmers who crossed them with 'better' Kerry Hill rams to produce the slaughter generation. Before 1914, a trade grew up in the sale of Kerry Hill rams to Shetland and elsewhere in Scotland for crossing. Draft ewes were sold to England for mating with Down rams. It is interesting how this one breed could serve as crossing and terminal sire and as a maternal breed, and the Kerry Hill is unusual in this respect.

The black face and leg markings were not a feature of the old breed of the Kerry Hills as described by Davies in 1810. Perhaps they came later, from the old stock of Eppynt Hill. In the 1920s and 1930s strong words were frequently used about these markings. Breeders were trying to promote the Kerry Hill as a crossing sire to produce cross-breds and the market preferred cross-breds to have coloured faces. There was therefore some commercial basis to the apparently purely aesthetic comments by T.A. Howson (sometime Secretary of the Royal Welsh Agricultural Society) writing in 1928:

> By the application of consistent and of careful breeding methods the dark, muddy
> and excessive markings have been, nowadays, converted into crisp and clean cut
> black and white, but, unfortunately, at the moment, there is a tendency to . . . breed
> sheep so utterly devoid of markings as to be practically white in face and legs . . . No
> matter how good an excessively white sheep may be . . . its place is certainly at
> home, and it should be heavily penalised in the showyard.

The Hill Radnor is a small sheep, generally resembling the Welsh Mountain but rather larger. Rams are usually horned, females hornless, the nose is grey and face and legs are brown and free of wool. While an attempt was made to form a breed society in 1926 this only bore fruit in 1955 when the Flock Book was started.

This breed was described by Trow-Smith in 1959 as having developed 'after many vicissitudes' from the sheep which Davis saw on the Kerry Hills early in the nineteenth century. Perhaps Trow-Smith's cautious pronouncement was a response to T.A. Howson's vigorous denunciation in Volume 1 of the Flock Book of the 'transparent absurdity of the mere suggestion' that the Shropshire was involved in the development of the breed. Howson maintained that the Hill Radnor is purely a type or offshoot of the Welsh Mountain. However, the story of this local breed must be more complex than that, and remains to be described authoritatively. Clark, who reported on Radnorshire agriculture in 1794, described how the sheep of the area 'thrived in England' on account of their hardiness and disease resistance. The Hill Radnor has rather finer wool than the Welsh Mountain, and while the two breeds must be closely related, it is difficult to see how the former can have arisen from the latter without some crossing with a third, as yet unknown, breed or race.

Kerry Hill champion sheep, complete with rosettes.

Champion ewe lambs, from the Honddu flock of Garthbrengy, Brecon, which won the Clun Forest Flock Competition in 1981. Bred by I.T. Davies & Son.

The Clun Forest is in a sense a hardy kind of Shropshire, and is rather smaller but of a less compact conformation. With the decline in arable sheep before 1914, the Clun Forest began to expand dramatically in numbers. Lowland flocks in England which did not breed their own replacement ewes (called 'flying flocks') found the Clun Forest an excellent and easily obtained sheep with a reputation for high fertility. Furthermore the transition from flying flock to permanent flock could be made easily when this breed was adopted because the flock could breed its own replacements which was not possible with cross-bred sheep.

The Clun Forest has always had a reputation for early fertility, ewe lambs being put to the ram when seven months old. Individual ewes have been famous for their prolificacy; in 1953 a three year old ewe produced quintuplets, having produced twins in 1951 and triplets in 1952. It is at least possible that this fecundity is due to selection by the early breeders; the fattest lambs, which were mostly singles, were sold for meat during the summer so the flock replacements tended to be the undersized lambs which would be twins, themselves more likely to produce multiple births.

In the heyday of the Clun Forest in the late 1940s and 1950s, up to 15 000 sheep of this and related types were sold on each of the four days of the Knighton September sales. Today the Clun Forest is in competition with the cross-breds, notably the Mule. One way breeders are meeting this challenge is by crossing Clun Forest ewes with Border Leicester rams to produce the English Halfbred.

Also noted for prolificacy is the Llanwenog, which arose in the valley of the river Teifi, north of Carmarthen, probably by crossing Shropshire rams with local dark-faced stock in the late nineteenth century. It has, if anything, a higher reputation for prolificacy than the Clun Forest. Today it is a neat, rather small, polled sheep with black face and legs and, like the Clun Forest, it has a tuft of wool on the forehead.

'Polled Welsh ram.' Youatt (1837) described several races of Welsh Mountain sheep and while a polled variety was particularly characteristic of Anglesey it was also seen in the central highlands. In the present day Hill Radnor, polledness is widespread in the rams and practically universal in the ewes.

Wiltshire Horn

Amongst British improved breeds of sheep the Wiltshire Horn is unique in having a very short fleece with no wool. In summer the hairy winter coat is shed, so the sheep do not require shearing. The origin of the breed is unknown although it is probably, at least in part, a relic of the Old Wiltshire Horn which was repeatedly interbred with other breeds, notably the Southdown, at the end of the eighteenth century. Perhaps the lack of wool was introduced at an earlier time by an unrecorded crossing with a foreign breed, for many southern and western Asiatic breeds of sheep are characterized by a very short fleece. Trow-Smith (1959) suggested that the Wiltshire Horn is a relic from a remote Roman ancestry, but this explanation does not seem very plausible.

During the eighteenth century the downs of Wiltshire and Hampshire were grazed by these distinctive, large, long-legged sheep which were essential providers of manure on the poor, chalky soils. They spread and replaced the native breeds of Hertfordshire and the North Downs. The Old Wiltshire Horn and the rather low intensity husbandry of which it was a part were, however, soon to become obvious victims for the new Southdown. Higher stocking rates and greater profits were possible with the new breed. The Wiltshire Horn was now, in Trow-Smith's words 'chased out of all parts of the country except the Vale of White Horse' by the Southdown. Even in its last retreat it was not safe and by 1819 it had all but died out, leaving a few relics of more or less outcrossed sheep in neighbouring counties. However, a new Wiltshire Horn was to stage a resurgence, showing how a breed may be written off if it

Wiltshire Horn, painted by Shiels. According to Low this breed only survived on one farm in its native county, at Hindon, in 1845. The farm had been bequeathed on condition that the proprietor kept a pure flock of the breed. As today, the fleece was only 1 kg in weight and the sheep was bare on the belly.

Wiltshire Horn ram, and flock, on the farm of Mr D.D. Lyn Davies at Blisworth, Northamptonshire, 1988.

does not conform to farming requirements at a particular time yet may still have a function to perform which may be unsuspected at the time of its demise.

H.J. Elwes was interested in the Wiltshire Horn and in the early 1900s interviewed breeders in the Midlands who had discovered its high value for crossing with 'Welsh Scotch and Down' ewes. By 1920 the Wiltshire Horn was most common in Anglesey and Northamptonshire, the latter county traditionally having a strong trade in old Welsh ewes. The breed society was formed in 1923 and while the breed has suffered from competition with other terminal sires it has retained its uses in cross-breeding.

Ryeland

In the closing decades of the eighteenth century the price for mutton was rising continuously while that of wool was stationary, except when the Napoleonic Wars interfered with the import of fine wools. The finest English wool came from the Ryeland sheep of Herefordshire.

The history of the Ryeland demonstrates clearly how the market can dictate changes in the characteristics of a breed and how valuable ancestral attributes can be lost as a result. By the middle of the nineteenth century the Ryeland was no longer a fine-woolled heath sheep, but a mutton-producing Down sheep, and today it is a specialized terminal sire intermediate between the Southdown and the Dorset Down.

In 1779 Ryeland wool was worth 30 pence a pound (0.45 kg) and the clip was between one and two pounds (0.45 to 0.9 kg). The coarser wool of the Lincoln was worth 5 to 6 pence a pound but the clip was at least ten pounds (4.5 kg). The Lincoln Longwool was also much more valuable for mutton and it could be integrated with arable husbandry because of its large size.

Progressive farmers began to change the breed and Dorset, Southdown, and New Leicester crosses all played their part in increasing carcass weight and coarsening the fleece of the Ryeland, while raising the weight of the clip. New husbandry systems involving turnip feeding had the same effects. Previously, Ryeland sheep were housed at night in 'cots' where

Shiels depicted sheep from one of the few remaining pure Ryeland flocks, belonging to Mr Tomkins of Kingspion, the celebrated breeder of Hereford cattle.

they were fed straw, and their dung was collected and distributed on the arable fields. The new husbandry system involved folding on roots and on rich grasslands. In response to market changes, the husbandry system and the breed had changed dramatically.

At the time some farmers at least knew where these developments would lead. Under the influence of Joseph Banks, King George III kept a Ryeland flock on heath and bracken land at Windsor, with a view to preserving the fineness of fleece, and attempts were made to cross the sheep with Merinos, but this project was soon abandoned.

The general conformation of the modern Ryeland is that of a large Southdown, but the white face indicates its separate identity. Indeed, it seems that crosses with Leicester, Lincoln, and Cotswold were still being made in the 1860s. A few breeders, however, paid some attention to the purer strains of Ryeland; at the time there was strong pressure from the Shropshire, and in 1903 a mere fifteen flocks of Ryeland sheep remained active. The first flock book was published in 1909 and it was closed to foundation stock in 1918. By 1920 there were eighty Ryeland flocks.

In 1974, 980 Ryeland ewes were enumerated, and in 1979, 1332. In 1986 it ceased to be officially listed as a rare breed, but its status as a commercial sheep is still not assured.

Youatt's illustration of 1837 shows an animal with the same general features as those illustrated by Shiels but with a conformation closer to the nineteenth century butcher's ideal. Today's Ryeland is a low-set, blocky meat sheep with a heavily-woolled head and white legs and face.

15   Downland sheep

The downland breeds of Oxford, Hampshire, and Dorset, together with the Shropshire, were major mutton-producing sheep in nineteenth century England, but of course they were also shorn for their wool. A feature of the new industry of more intensive breeding of sheep was the long distances that the animals were moved over the country to satisfy the new urban markets and this inevitably meant the breakdown of the old regional types. From the beginning of the nineteenth century the fashionable Southdown was increasingly interbred with all the downland breeds and this transformed them. The Southdown was also interbred with the unimproved Norfolk, a sheep that was much disliked by Arthur Young, and this cross was ancestral to the modern Suffolk.

A recent article by Bowie (1987) has reviewed the changes in the downland sheep from 1792 to 1879, a period that was crucial for the development of all breeds of sheep in Britain. Youatt described the Old Hampshire of the opening years of the nineteenth century as 'like the Old Wiltshire, horned, tall, light, and narrow in the carcass, with a white face and shanks'. This breed was extinct by the time Youatt was writing in 1837, as was the mainly black-faced Berkshire Nott, both of these being ancestral to the new downland breeds which were all products of Southdown crosses.

Improvement of the Southdown was carried out first by the celebrated John Ellman of Glynde and it soon enjoyed the patronage of the great landowners, in addition to being given much publicity in the writings of Marshall and Arthur Young between 1770 and 1813. Again, as quoted by Bowie, the Southdown was 'in favour with gentlemen farming their own estates, for the finer quality of the mutton'.

Perhaps the most successful of the new breeds, at least initially, was the Shropshire, which was developed from Southdown crosses on the local breeds of the Welsh border.

Southdown

In 1789 Gilbert White of Selborne described polled, black-faced sheep with speckled legs and fine wool, on the eastern part of the South Downs. It was about this time that John Ellman of Glynde, near Lewes, began improving his stock. Little is known about his methods but he is understood to have mated ewes with undesirable characteristics to rams selected to correct the weakness. This could mean he tended not to cull ewes, and this, in the light of modern knowledge, would have slowed the rate of improvement. In the 1780s and 1790s many illus-

F.W. Keyl (1823–1873) painted a Southdown flock in its accustomed habitat (above); from the Royal Collection.

Left: Jonas Webb of Babraham, near Cambridge, was another famous breeder whose father had experimented in Norfolk with the improvement of the native sheep by Southdown crosses. Sussex stock was used to found the Babraham flock, which provided very many prizewinners as well as helping to establish the breed in France, by arousing the interest of the Emperor Napoleon III. There was great competition between Sussex and Cambridgeshire Southdowns; the Sussex breeders 'declared that they could get as good legs of mutton as Webb did but the Babraham shoulder was beyond them'. Lithograph by C. Hullmandel after J.W. Giles (1842).

trious names became associated with the Southdown, largely due to the missionary zeal of Arthur Young. The Dukes of Bedford and of Grafton and Thomas Coke of Holkham started Southdown flocks and there was much competition with the New Leicester. Ellman did very well from the new breed; in 1802 a record was set when he hired a ram for two seasons to the Duke of Bedford for 300 guineas.

By the time Ellman retired in 1829, the Southdown had an enthusiastic following among the fashionable, and an established reputation for quality of mutton. However, accidents did happen. One anecdote records:

> . . . the Duke of Bedford was once, when dining with Lord Sligo, earnestly recommended to taste a fine haunch of Glynde mutton . . . His Grace readily acceded, but no politeness to his noble host could induce him to finish the slice, or to say it was otherwise than rank in flavour, and terribly tough! On enquiry, the disappointed Marquis ascertained that his shepherd, who had been ordered to kill the 'best' Southdown sheep, had actually slaughtered a ram for which Lord Sligo, a few weeks before, paid Mr Ellman 200 guineas.

The Southdown rapidly became the premier sheep of the English downlands. At first Lewes Fair was the main market and in 1793, 30 000 sheep were sold there. Throughout the nineteenth century the Southdown was the central element on every Sussex hill farm. Up to World

War I the rest of the farming operations were geared to the needs of the Southdown flock. The system was a combination of close folding on arable crops at night, and wide ranging on downland pasture in the daytime. Lambs were sold off the farm for fattening on richer land. A ram trade was also built up, Southdowns being found very suitable for crossing with Romney Marsh sheep in particular, producing lambs with remarkably light bone. Findon Fair is now the biggest market for Southdown rams, but numbers are much lower nowadays. Indeed the decline in the Southdown sheep husbandry system started because dairy farming was more profitable, and while there were 296 138 sheep in Sussex in 1926 there were only half that number in 1954.

The Southdown is probably the sheep equivalent of the nineteenth century Shorthorn. Firmly associated, initially at least, with a particular part of the country and with a few illustrious breeders, each was brought (so its supporters claimed) to a pitch of perfection in its own right. Each made an enormous contribution to the formation of other breeds and to the development of livestock industries overseas, and was favoured by the nobility and gentry, many of whom paid very high prices for pedigree stock.

Shropshire

The sheep of the Shropshire heathlands had long been known as hardy, fine-woolled but slow maturing animals, and in the early years of the nineteenth century George Adney and the Meire family, both of the Severn valley, set about improving the stock. Southdown and perhaps New Leicester rams were used to improve carcass conformation, suppress horn growth, and produce a darker face. Considerable effort was also devoted to the task, perhaps less apparently rewarding at first, of improvement by selection within the original stock. When the Royal Show came to Shrewsbury in 1845 the emerging breed did not fit into any of the categories of Leicesters, Southdowns, Longwools, or Mountain Breeds, showing that even by this rather early date the breeders had succeeded in developing a new type of sheep.

The new breed was not merely suited to local conditions. By 1882 when the Flock Book was established (the first in Britain) flocks were in existence all over the United Kingdom. Many of the breeders profited from the hiring or sale of rams for crossing with other breeds for the production of meat lambs.

The fame of the Shropshire spread overseas and by 1911 it could be described (in Wright's Cyclopedia) as 'the most ubiquitous sheep extant', finding favour in the USA, Canada, South

Three Shropshire wethers owned by Mr John Coxon of Freeford Farm, Lichfield, which won first prize of £15 and a silver medal in the Smithfield Show of 1862. The flock was founded in 1845 and produced many prizewinners; from 1860 the Shropshire had its own class at the Royal Show. Painted by Ralph Whitford (1863).

Royal Shrewsbury 2nd from the 1915 Flock Book. For no very clear reason overseas buyers, especially North Americans, favoured the heavily-woolled head, which was quite inappropriate to British conditions and was a major cause of the breed's rejection during the 1920s and 1930s. Fortunately some breeders had held fast to the older, clean-faced strain.

The 1982 Shropshire Flock Book Centenary, with Mr Ben Dew, owner of the largest flock of Shropshires in Britain, in the foreground.

America, Russia, France, Germany, Australia, New Zealand, South Africa, Jamaica, and the Falkland Islands. At that time the American Shropshire Registry Association was the largest livestock organization in the world.

Disaster struck when World War I and the 1923–4 foot-and-mouth outbreak cut the Shropshire off from the foreign market for by this time, breeders had concentrated so heavily upon export requirements (whose most obvious physical expression was the heavily-woolled head) that there was no home market for Shropshire rams. The breed society revised the breed description in 1941 to favour wool-free heads but the Shropshire continued its numerical decline and by 1972 there were only ten registered flocks.

Since then, the Shropshire has effectively advertised its own distinct merits. It has enlarged its band of devotees with 36 new flocks being added between 1974 and 1980. The class at the Royal Show, which was discontinued in 1963, was re-established in 1977, but the breed will have to fight hard to regain the ground which it has lost to the Suffolk.

Oxford Down

The Cotswold (see p. 143), under the influence of the New Leicester, had become by the 1840s a big, long-woolled sheep and the most important breed in Oxfordshire. Farm sales advertisements show that after this initial expansion the range of the breed began to contract and the reason for this was competition from a new breed – the Oxford Down. There had been room for improvement in the Cotswold, most notably in the quality of its meat, and this was started by the introduction of Southdown blood. Mr John Talmadge Twynam is credited with the early development of the Oxford Down; he further crossed Cotswold rams with Hampshire ewes, starting in 1829, to give cross-bred sheep whose type was fixed by selection, particularly by breeders in the Witney area.

By the 1850s Oxford Down rams were being sold in large numbers (first going to the USA in 1853). It took rather longer for the value of these rams for crossing with the cross-bred ewes of the Scottish borders to be recognized; the first batch of Oxford Downs was sent to the Kelso ram sales in 1884. At first the Oxford had practically no rivals for this purpose but by 1900 the Suffolk had begun to make inroads, acquiring a far better reputation for early maturity and leanness and for the production of the rather small lambs that were then in

'The glory of the county – the most profitable sheep to the producer, the butcher and the consumer.' Thus wrote C.S. Read in his essay on the agriculture of Oxfordshire (1854). He was a particularly eloquent Oxford Down enthusiast. The breed certainly became a prominent part of the farming of the county, appearing on this £5 note issued by the Witney Bank in 1913.

demand. From sales of up to 1000 rams a year at Kelso in the 1930s and 1940s the Oxford Down crashed in popularity; in 1962, 230 were sold at Kelso, but within ten years it was very much a minority breed with fewer than 1000 registered ewes. Once the Suffolk had taken the bulk of the terminal sire trade, the Oxford Down breeders found it impossible to cover their costs, especially as, in their native areas, the keeping of flocks did not fit in at all with spring barley cultivations, and many flocks were dispersed.

The trend has now been reversed, and the breed is again finding favour as a terminal sire. In 1985, 85 were sold at Kelso, and trials have clearly shown the breed's merits as sires of heavy, early yet lean lamb. In 1985, 18 new flocks were formed, bringing the total to 97 flocks in 43 counties.

Early on, the Oxford Down's combination of high quality meat, and quantity of wool (attributed respectively to the Southdown and New Leicester components of its ancestry) found favour with agricultural experts such as Philip Pusey. The breed was very frequently successful at fatstock shows when its main importance was as a pure-bred arable sheep. Indeed, success at Royal Shows was a criterion for the retrospective registration of rams. Whitford painted John Treadwell's flock around 1860–70.

Norfolk Horn and Suffolk

Black-faced and horned in both sexes; many writers have remarked on the similarity of the Norfolk Horn in these respects to the Linton or Blackface hill sheep of the Pennines, but a historical connection has yet to be found. This painting by Shiels is of a Norfolk ewe with her hornless Southdown-cross lamb.

The native breed of Norfolk was well known for its tasty, lean mutton but this did not compensate for its 'wretched' and 'contemptible' appearance, in the words of Young (1771). The black face and legs, horns in both sexes, lack of compactness of carcass and refusal (typical of the heathland sheep) to be confined by fences, were not compensated by its prolificacy and the high quality of its wool. None the less at the end of the eighteenth century there were still some very large flocks of these sheep on the light soils of Norfolk, Suffolk, and Cambridgeshire.

Early attempts at improvement involved crosses, principally with the Southdown, old Wiltshire Horn, and Lincoln. Young himself used the Southdown and by 1791 had 350 Southdown × Norfolk ewes. The infusion of Southdown blood, creating a better meat sheep, was timely as it coincided with the spread of turnip husbandry and the developing practice, so effective on light land, of co-ordinating sheep-farming with corn-growing.

By 1845 Low reported that the 'perfectly pure Norfolk breed' was 'rare', and the new Suffolk breed was first recognized at the Suffolk Agricultural Association's show in 1859. Early pioneers of the breed, not forgetting the part played by Young, were George Dobito of Cropley Grove, and the Garretts, agricultural engineers of Leiston. With this firm base of local support the breed could weather such criticisms as those made at the Royal Show of 1867 at Bury St. Edmunds, of the breed's 'long legs, short ribs, thin necks, bare backs and naked heads'.

In 1886 the Royal Show was due to be held in Norwich and the Suffolk Sheep Society was formed early that year to make sure that Suffolk sheep of a uniform appearance were entered for the 'Suffolk black-faced sheep' prize classes. By the end of the century, rams were being

Norfolk Horn sheep at the Cotswold Farm Park.

sold in high numbers to Scottish sheep farmers and the use of the Suffolk ram as a terminal sire was being established.

In contrast by 1900 there were probably only 300 Norfolk Horns still in existence. Despite the lack of numbers, some breeders were interested in the potential of these sheep and in 1902 a Norfolk carcass gained first prize at Smithfield Show. The most important flock was that of James Sayer at Lackford, Bury St. Edmunds, started in 1895 which after 1919 was the only surviving flock. Mr Sayer was also a prominent breeder of Suffolks but he always kept the two flocks separate. Inbreeding of the Norfolks was, however, inevitable and led to a decline, so that there were a mere twenty ewes in 1930. In 1950 there were only eight ewes and five rams and in 1959 after Mr Sayer's death the last survivors were given to the Zoological Society of London to be kept at Whipsnade Zoo. In the late 1960s cross-breeding with Suffolks was resorted to, and the flock passed into the ownership of the Royal Agricultural Society of England. In 1973 there was one infertile pure-bred ewe, one pure-bred ram, and twenty-eight cross-breds. The Rare Breeds Survival Trust provided registration facilities and support for the relic, which it designated 'New Norfolk Horns', but the breed did not conform to the Trust's criteria for acceptance because of the percentage of blood attributable to other breeds. However, in 1986 the percentage of Norfolk Horn blood was high enough, following the implementation of a controlled mating plan, for the Trust to accept the breed and for it to be renamed Norfolk Horn. Today there are ten flocks and a total of about 400 sheep, and in 1982 Norfolk Horn sheep were successful at Smithfield once again.

Many of the characteristics of the Suffolk recall its Norfolk origins, notably the black face, and a tendency, still not entirely banished, to produce sheep with horns. The leanness of Suffolk meat has also been attributed to the Norfolk. The old breed was also notorious for a tendency to scrapie (an incurable wasting disease, transmitted as though it were an inherited defect) but the degree to which the Suffolk has this tendency in comparison with other breeds is not known.

Early on, flocks of Suffolks were established in Ireland, Scotland, and Wales (1891, 1895, and 1901 respectively). Exports were also recorded in the 1890s to Europe, Russia, North and South America, and elsewhere. After 1918 the Suffolk's ascendancy as the terminal sire breed for mating with cross-bred ewes became very marked. Indeed, the female progeny of such crosses was soon found to have distinct advantages as a maternal sheep. Hammond pointed out in 1947 how the only other breed approaching the Suffolk in the combination of early maturity, fertility, and leanness of meat necessary for this form of cross-breeding to succeed was the Dorset Horn. It was at this time that the value of female lambs as breeding stock outweighed their slaughter value, a result of the appalling weather of 1946 and 1947 when more than 20 per cent of Britain's breeding ewes died.

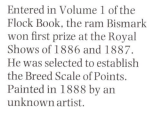

Entered in Volume 1 of the Flock Book, the ram Bismark won first prize at the Royal Shows of 1886 and 1887. He was selected to establish the Breed Scale of Points. Painted in 1888 by an unknown artist.

The Suffolk has been widely exported. In 1970 the record number of 1269 went abroad. It is also popular in Britain because of its rather early breeding season. Indeed, early lambing is general practice in flocks which rear ram lambs for sale for breeding, as the rams can be used for breeding in the autumn of the year of birth.

Today about 50 per cent of the genes of British lamb production come from the Suffolk and 10 per cent from other Down breeds. The rest of the genetic endowment of the slaughter generation comes from hill breeds and crossing sires. The correct choice of terminal sire is very important for the production of high quality lean lamb. While the Suffolk is the most popular terminal sire breed in Britain today, there is much competition, notably from the Texel. Numerically weaker breeds like the Oxford Down, Southdown, Dorset Down, and Hampshire Down are also considered suited to particular lamb production systems and some farmers use cross-bred rams like the Suffolk × Texel for siring the slaughter generation.

Suffolk-type sheep near Woburn Sands, by Sylvia Frattini (1982).

Hampshire Down and Dorset Down

Three ewes in a landscape; first prize pen of Hampshire Down ewes, Smithfield Club Show 1879, painted by Ralph Whitford. Today the Hampshire Down whether pure-bred or cross-bred is very successful in carcass competitions; it is one of the largest bodied of the early-maturing breeds.

Southdowns had been adopted by many fashionable and progressive farmers in Wiltshire, Hampshire, and Berkshire in the first few decades of the nineteenth century. They had much better carcass quality than the local races, namely the Old Hampshire (tall, horned, narrow in the carcass with white face and legs), the Wiltshire Horn (tall and close-horned), and the Berkshire Nott (like the Wiltshire Horn but dark-faced and sometimes polled).

Between 1800 and 1860 the downland sheep walks were being ploughed for cereal crops and the system of sheep folding came into use. As artificial manures and new forage crops became generally adopted the advantages of selling meat in the form of lambs rather than mutton became apparent. Farmers crossed Southdowns with local breeds and the result was a rapid growing, dark-faced and hornless sheep with wool over the poll and forehead. The heavy head and wide shoulders were also notable features. Other characteristics were high carcass quality and early maturity, which meant that wether lambs of the year could be sold

Two Dorset scenes painted in the 1930s by William Gunning King; Dorset Down sheep in early spring (right) and at shearing time (left).

in the autumn as high priced lamb. Indeed, in 1855, farm accounts show that lambs were sold off the farm in May and June.

The new Hampshire cross-bred was much better fitted to this kind of husbandry than the old breeds, and William Humphrey, who farmed near Newbury in Berkshire, is credited with the further development from 1842 using Jonas Webb's Cambridgeshire Southdowns. Humphrey attended the first Royal Show and this is presumably where he met Webb's sheep.

With his followers, Humphrey developed a fine meat-sheep which was recognized by the Royal Agricultural Society in 1861. The flock book and breed society were started in 1890. By this time the husbandry system of which sheep manure had been an essential part was on the way out because cereal crops on downland could be produced much more cheaply by using artificial fertilizers. At the same time, dairy farming, particularly after 1914, came into competition for the land, and the loss of shepherds during and after World War I was one of the most telling blows. The big downland sheep fairs of Overton and Stockbridge ended in the 1930s, although Alresford Fair lingered on in a changed form until 1954.

In 1855 the improved Hampshire Down was first exported to the USA, and in the 1920s American registrations of the breed were running at 20000 annually. However, the Hampshire Down breeders did not fall into the same trap as the Shropshire breeders, that is allowing the fashionable American market to dictate the form of the breed to the exclusion of commercially useful characteristics. Indeed, these overseas populations of Hampshire Downs have not been resorted to for 'new blood' for the British population of the breed, unlike other British sheep (notably the Southdown, and the Dorset Horn/Poll Dorset). The breed is now distributed world-wide and there are around one hundred flocks in Britain.

Trow-Smith (1959) stated that the Southdown was also used to improve the 'local polled cousin' of the Dorset Horn. Later, Hampshire Down blood was introduced. This led to the Dorset Down, with a breed society which was formed in 1906. Prominent among the early breeders in the 1840s was Saunders of Watercombe, near Dorchester, who exchanged rams with Humphrey, the Hampshire Down pioneer. Today there are about one hundred flocks. Both these breeds are terminal sires conferring early maturity on their cross-bred lambs.

16   Goats

Although the meat of goats has never been popular with the British, their milk has always been valued, their fat has been used for tallow, their skin provides the best parchment, and their undercoat the finest wools. Until the recent boom in cashmere and angora, goat wool had always been considered inferior to that of sheep, but today the demand for these, the finest of all wools, far exceeds the supply.

The value of cashmere was realized as soon as European travellers began to bring back fabrics from Tibet and neighbouring countries, the original home of the Kashmir goat. Low, writing in 1845, described how efforts were made to import these goats to England but they were not successful. Low claimed that in his time goats were decreasing in numbers throughout Britain, having been present in large numbers in the Highlands of Scotland where they had been milked for cheese-making as also in Wales. He stated, however, that there were still a great many goats in Ireland.

The British Goat Society was founded in 1879 and opened a Herd Book in 1886. This was to record the best milkers irrespective of breed. Later, separate sections were opened for the distinct breeds, although cross-bred goats were still included, and in 1923 the name 'British' was approved by the Goat Society to cover all the cross-bred milkers. In modern times the goat has increased in popularity as a provider of milk, particularly now for making yoghurt.

The standard work on the goat is still *The Book of the Goat* by Pegler, first published in 1885 but running into many editions up to the 1930s. The majority of pure-bred goats at the present time in Britain are of Swiss origin, the original stock having been imported at the beginning of this century.

The sections of the British Goat Society Herd Book are as follows: British (opened 1923 for cross-breds); Toggenburg (1905); Nubian and Anglo-Nubian (1910); Saanen (formerly Swiss, 1922); English (1921); British Alpine, British Saanen, and British Toggenburg (1925); Golden Guernsey (1922 in its own herd book). The four main breeds of milk goats today are the Anglo-Nubian, British Saanen, British Alpine, and British Toggenburg. Then there are the cross-bred British goats and the Golden Guernsey and the English Guernsey, which are steadily increasing in popularity.

Shiels' vigorous depiction of Welsh goats was not included in Low's book (1842), perhaps because this breed was beyond the fringes of respectable farming.

Anglo -Nubian

This breed of goat was imported into Britain in the middle of the last century from the Middle East and India. The goats had been specially chosen to provide milk on board ship during the journeys from Asia to England. There have been no new imports for the last fifty years but the breed has remained stable and still has an 'eastern' appearance. The goats are coloured, have a short glossy coat and long lop ears. The milk has a higher butterfat content than that of any other breed.

Alpine breeds

The wild goat, *Capra aegagrus*, from which all domesticated goats are derived is a mountain ungulate. Perhaps this is why the goats of Switzerland are superior to those of the lowland countries in Europe, the Swiss environment being closest in kind to that in which they evolved. They have been imported into Britain since the first decades of this century.

The best known of the Swiss breeds is the Saanen which has a world-wide reputation as a high milk-producer. It comes originally from the Saanen and Simmental valleys of the canton of Berne. The Saanen and British Saanen are white goats that are hornless and of slender build.

The Toggenburg in Switzerland is close to the Saanen in performance. In Britain the breed can be either horned or hornless and is fawn or light brown with white markings. The hair on the back and hind legs can be long in both sexes but this is less common in England, particularly in the females.

The British Alpine breed is usually hornless. It is a product of cross-breeding and has the appearance of a black and white Toggenburg. It should have a short, glossy black coat with white lower limbs.

These alpine breeds are notable for their 'Swiss markings'. These are defined by the British Goat Society as 'white facial stripes from above the eyes to the muzzle and edge and underside of the ears. White legs below the knees and hocks, also white on the rump and on and about the tail.'

Golden Guernsey

Angora goats produce mohair, an extremely valuable fibre. In 1983 Britain imported it to the value of £36 million. Remarkably high prices have been paid recently for Angora breeding stock.

This is a breed that is new to England and it is not mentioned by Pegler (1885 or later editions). It originated in Guernsey and was mentioned in a guide book to the island in 1826. However, it was not exported to the mainland of Britain until 1965 and in 1967 the English Golden Guernsey Club was formed. These goats have a beautiful golden colour to their skin and hair, which is long and silky. They have no Swiss markings. Their appearance (see illustration on p. 92) is their major asset, as milk yield is not high. The Golden Guernsey and the Bagot are the only British breeds defined as rare.

English, Welsh, and Irish goats

The goat is unlike any other species of livestock in Britain because it never received the attentions of eighteenth century livestock improvers. This was, presumably, because goats were looked on either as household animals, 'the poor man's cow', or as game in the same category as deer. Culley (1807) wrote:

> Of goats and deer I know very little; but suppose that the different species of those animals might be greatly improved, by the simple and plain rule of selecting the best males and best females; and breeding from these, in preference to the promiscuous methods, which at present, I am told, are too much pursued.

During the 1920s an effort was made to conserve and improve the 'Old English Goat'. This may be an ancient breed but there is no strong evidence that it has survived introgression from the Swiss cross-breds. Pegler (1885 and later editions) describes the Old English Goat as coloured or white with long hair and horns set wide apart; this is a fairly general description. Of the Welsh goat he wrote merely that it resembled the mountain goat, but the Irish goat he claimed was quite different, being coloured, with long shaggy hair, a long ugly head, and large pointed horns.

The only documented breed of goat to have survived in Britain since the Middle Ages is the Bagot goat and this may have originated in Switzerland.

Feral goats

Since Medieval times herds of goats have been pastured on hills, mountains and moorland. Many goats, over the centuries, have wandered from their owners' control and have flourished and bred as wild animals. This has resulted in numerous small, widely separated herds of mixed origin in many of the upland areas of Britain. These herds are living relics of primitive domestic livestock and they preserve a gene bank that is probably irreplaceable and which may be of commercial value.

Recently, it has been found that some feral goats have an undercoat of wool that is fine enough for the production of cashmere, the most highly valued wool in the world. Experiments are being carried out to determine whether a viable industry can be developed by cross-breeding feral goats with domestic stock to produce marketable quantities of cashmere. This would be an ideal use for Britain's feral goats as long as the herds are not depleted to the point where they are in danger of extinction. At present there are about 150 000 domestic goats in Britain and 3000 feral.

Feral goats near Lynton, Exmoor. British feral goat herds differ greatly in their history. Some have been left to their own devices for up to two hundred years; others including these are of more recent establishment.

Bagot goat

The Beggar's Oak – Bagot's Park.

In some ways the Bagot goat is in the same position in the large mammal fauna of Britain as the Chillingham cattle and the continuity of its history is equally fascinating. This breed of goat is descended from domestic stock that had reportedly lived wild for 600 years in Bagot's Park which covered 800 acres (324 hectares) at Blithfield in Staffordshire.

Bagot goats are unlike any other British goats in appearance but they closely resemble the Swiss Schwarzhal breed. The head, neck and shoulders should be jet black, the rest of the body is pure white, and the hair is long and shaggy. The males have horns that are large and spreading at their tips, while those of the females are smaller and set close together.

The history of this ancient breed and some of the ideas about their origin have been described by Whitehead (1972) who recounts the belief of the late Lord Bagot that a number of Schwarzhal goats were brought to his estate from the Valley of the Rhône by Sir John Bagot on his return from a Crusade in 1387. Lord Bagot died in 1962 and some of the goats were sold; in 1979 the remainder were handed over to the Rare Breeds Survival Trust and they are now distributed among breeding centres.

The Beggar's Oak, Bagot's Park, Staffordshire, with Bagot goats.

The Bagot goat and Dexter cattle are the only breeds of British livestock to bear a family name. The long association of these goats with the ancient Bagot family is only paralleled by that between the Chillingham cattle and the family of the Earls of Tankerville. The tomb of Richard Bagot (1552–96) is surmounted by this goat's head.

Bagot nanny, Bemborough Ann, with Mr Joe Henson, Chairman of the Rare Breeds Survival Trust and a pioneer in this branch of genetic conservation. Bagot goats represent a considerable challenge to the breeder. Although until recently they lived as wild animals in Bagot's Park, their ancestral domain, they are not notably hardy animals particularly when young.

17　Pigs

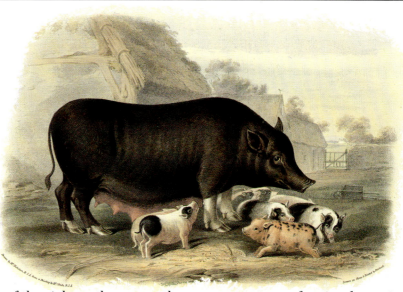

This painting (left), by an unknown artist, was reproduced as an engraving in Pitt's survey of Staffordshire (1794). The 'spherical' pig had a live weight of 802 lb (361 kg) and was killed at the age of two and a half years. It belonged to Mr Dyott of Freeford Manor, Lichfield.

This picture by Shiels represents a sow of the Siamese breed, imported from Singapore, with her litter by a halfbred Chinese male. The mixed parentage has led to a tremendous diversity of colour. The neatly rounded form of the piglets shows the early-maturing characteristics inherited through Far Eastern blood, which made pigs of this type so successful as porkers.

Up to the middle of the eighteenth century there were two types of native domestic pig in the British Isles; a very small, dark-coloured pig with prick ears that was found principally in the Highlands and Islands of Scotland, and a much larger kind which could be white or dark, often with lop ears, that was found throughout England. The very small pigs lived almost wild and were apparently left to forage on hills and moorland on their own, feeding on their natural diet of roots, other vegetation, and animal matter. On the sea shore the pigs scavenged for shellfish and other food, while they would wreak havoc in cornfields if they could get into them.

The larger breeds of pig, as happened with other livestock, had evolved in the different districts of Britain into separate breeds; the Yorkshire, Berkshire, East Anglian and so on. By the middle of the eighteenth century, however, a dramatic change was beginning to take place with the importation of Chinese, Siamese (Tonkey) and Neapolitan pigs. These pigs were considerably smaller and fatter than the old British swine and they had the dished face that is now common on almost all breeds. Before the import of Asian pigs the native breeds had a facial profile that was nearly straight, as in the wild boar. The very interesting models by George Garrard sculptured in about 1800 depict very clearly the conformation and size of the pigs of the time (Clutton-Brock, 1976).

Following his principle to improve all the livestock on his farm at Dishley, Bakewell in the 1780s attempted to produce a superior breed of pig by his practice of inbreeding. The contemporary comments on his results suggest that he was not entirely successful, but in general there are extremely few records of the early efforts to improve the British pig breeds. Both Youatt (1847), and Low (1845) gave accounts of the regional types of pigs throughout

A mezzotint published in 1802. The painting was by George Morland and it was engraved by William Ward in 1797.

Britain and of the imported Chinese which are described as being of two sorts, the white and the black, but by this period there can have been only a few of the unimproved native breeds still in existence.

With the widescale clearance of forests, the ancient practice of putting pigs out to pannage, that is allowing them to roam free in the forests to feed on acorns, roots, and beechmast, had dwindled by the middle of the eighteenth century but there were still great herds of swine in the New Forest and in other woodland areas. However, the pig industry was gaining strength in centres where they could be fattened on the waste from distilleries and the whey from cheese-making. Potatoes became an increasingly popular food for the fattening of pigs and Trow-Smith (1959) wrote that quite extensive areas of potatoes were grown at the end of the eighteenth century expressly for feeding pigs.

Pigs and their antics have given delight to artists for years as this selection by unknown painters (right) shows.

204

Although, as noted by Trow-Smith, there are few records of the beginnings of recognizably modern pig-farming, the nineteenth century saw advances in the design of pig housing and in the techniques of rearing and fattening, which were begun by the breweries.

The pig is a prolific breeder and a new generation can be produced every year, so that the intermingling of the breeds has been much more confused than with any other species of livestock. Experimental interbreeding during the nineteenth century was very extensive and no breed escaped from the Chinese or Neapolitan cross. Nevertheless, the established breeds of British pig today all have a lineage that has some ancient components, and if genetic diversity is to be maintained it is necessary that they should be conserved.

Heavy Sow by Ted Roocroft (1982). In holly wood (30 × 60 cm).

Berkshire

'Pig, bred by Mr Loud, Warwickshire.'
Five of the six famous white points of today's Berkshire, on the feet and the tail tip, are visible in Shiels' painting. The sixth, on the nose, is absent. The red colour is not associated with the Berkshire today, although russet spots were still frequently seen in the 1920s. Whether or not the red colour is an ancestral feature, derived from the ancient pigs of Britain, or whether it was introduced, is not known.

Today the Berkshire is a black, prick-eared breed with some white markings. It is an early-maturing pork breed. Two hundred years ago, a type of pig known as the Berkshire was coming into favour as a store pig, reared on the rich lands west of London and sold for fattening on the wastes of food processing industries, the slaughterhouses, bakeries, dairies, breweries, and distilleries of London.

Early development of the Berkshire probably owed a good deal to Chinese and Neapolitan pigs, as testified by the dished face and early maturity, but uniformity of type was, so far as is known, not approached until the second half of the nineteenth century. The early accounts of the breed are confusing in that sometimes 'Berkshire' refers to the pigs of the county, and sometimes to a general type. The breed was first accepted at the Royal Show in 1862 and the British Berkshire Society was founded in 1884.

Distinguished names were associated with Berkshires from early times. Lord Barrington's herd was influential in the 1820s, and noteworthy at the time for its black colour, which did not become a breed characteristic for several decades. Later on, Mr Heber Humfrey was one of the most important breeders and exporters, helping to found the Society and compiling the Herd Book for twenty years.

At the turn of the century the Berkshire was considered Britain's finest breed for pork and had been exported world wide. In carcass competitions at the Smithfield Show, of 46 first prizes offered in these classes between 1904 and 1917, 41 went to pure-bred Berkshires, and 2 to Berkshire crosses.

The Berkshire had come a long way from its humble beginnings. As with cattle and sheep, the development and refining of the breed had become a fashionable pursuit. Show points (like white feet and a white tip to the tail) were stressed and so was fat covering, although contemporary authors criticized the production of over-fat pigs as commercially irrelevant, indeed harmful. At the end of the nineteenth century the public demand was for bacon, preferably cured by the Wiltshire method, and British pig farmers were unable to supply the quantities of consistently good meat that were required. The Danes, who were heavily dependent on pig farming for foreign earnings, spotted the opportunity. British pig breeders failed to compete with the scientific approach adopted by the Danish pig industry and lost the market.

In this, pig farmers were no different from other British meat producers. The pig industry did not try to satisfy the home mass market. In the Victorian era of rising real wages and a free trade economy based on industry and commerce, cheap imports satisfied the home market. British markets had been opened to European fatstock in 1842 and in 1869 the carcass market at Smithfield obtained 25 per cent of its stocks from abroad. The same applied in the 1920s and 1930s, times of low productivity in farming and a period when the settled pattern of Victorian and Edwardian land tenure was broken up. In an era of free trade there was no national need (as there was during and after World War II) to be self-sufficient in food. Instead there was strong motivation to produce the world's best pedigree livestock, and this is why British breeds generally have shaped the western world's farm animals yet, in their native country on very many occasions, have given way to foreign imports of meat.

In today's highly intensive British pig industry the Berkshire has no place. In 1949, 196 boars of the breed were registered; in 1979, only 16. The breed has to be protected by special conservation measures if it is to survive. The Rare Breeds Survival Trust has financed reimportation of boars from Australia (for example, Lynjoleen Ambassador 1183rd in 1976 who sired nine notified pedigree litters in three herds) and pays incentives to owners of boars of rare blood lines. There is a small but expanding market for meats from rare breeds of pigs and it is the development of such ventures, coupled with scientifically planned measures, that are most likely to achieve conservation of a breed.

In the course of its history the Berkshire like the Southdown sheep and the Shorthorn cattle became the standard of fatstock perfection by which other breeds were judged. Very many were exported, but the trade for which the breed was suited, namely early-maturing pork, has contracted and with it the numbers of Berkshires, so it is now one of the rarest breeds. Of the 850 000 breeding sows in Britain today fewer than 150 are Berkshires. There are no details of the artist or subject of this picture.

British Saddleback

The Essex and Wessex Saddlebacks had distinct histories when they were merged into the British Saddleback in 1967. The Essex and the Berkshire are the only British pigs with a history of involvement by aristocratic breeders leading to commercially valuable breeds, though the precise degree of indebtedness of the twentieth century Essex to Lord Western's activities in the early nineteenth century is difficult to ascertain. Lord Western imported Neapolitan pigs, having been impressed by them while on the continent. He inbred them, possibly crossing them on to the local pigs and introducing Berkshire blood. A new breed, the Essex Half-Black, emerged. Further development led to an almost entirely black pig and even today the extent of the white belt in Saddlebacks is very variable. The Essex was always considered rather lighter in weight in the shoulder, and matured earlier than the Wessex.

The Wessex was a local sheeted black and white breed of unknown origin which was exported to the USA perhaps as early as 1820, and which gave rise there to the Hampshire. Like the Essex Pig Society, the breed society for Wessex pigs was founded in 1918. The Wessex had, as a breed, gone through a difficult period when the only herds were kept outdoors in the New Forest, but from the 1920s onwards growth in numbers was spectacular. For a long time the Wessex Saddleback was second only to the Large White in numbers of registrations but by 1964 it had slipped to fourth place, with 1732 registrations (Large White 22 528; British Landrace 15 091; Welsh 2713).

Layley & Malden claimed in 1935 that the Wessex was the 'only British breed free from Chinese influence', and was descended from the free ranging pigs of the New Forest. These statements need explanation. In 1917 they found only six boars they considered to have been bred pure, through two families in Plaitford and Landford. However, the foundation stock in Volume 1 of the herd book (1919) included 76 boars and 260 sows and this number included pigs with Chinese influence. Layley & Malden claimed that the Wessex Pig Society's insistence on 'Old English Baconer' type later weeded out this Chinese influence. This scraper-board illustration is by Tunnicliffe.

Saddlebacks generally have been closely identified with outdoor pig husbandry. In the old days this type of farming was not highly regarded, being associated with the production of store pigs of doubtful quality which were sold for fattening on swill. New systems have been developed, integrated with arable farming, particularly suited for well-drained land in central southern England. Sows farrow in huts, and piglets are weaned at three weeks and moved indoors for fattening. Experience has shown the British Saddleback to be hardy and well suited to the outdoor life, but in some ways it is not up to modern requirements because the subcutaneous fat which probably contributes to hardiness detracts from the value of the carcass.

Outdoor sows are often Landrace × Saddleback crosses in pig to a Large White to produce the slaughter generation. Some outdoor pig herds breed their own replacements but the trend is for specialist breeding companies to supply highly bred, disease-free gilts and new strains such as the Camborough Blue are now available which combine the merits of the Saddleback and the Landrace.

Belted pigs have been known for a very long time, being represented in Italian art of the fourteenth century. The Essex Pig Society was founded in 1918 and produced its first herd book the next year. The Society drew attention to the breed's resemblance in colour pattern to the Old English sow painted by Shiels.

Large White and Middle White

The Large White is the most widely distributed breed of pig in the world. Often known abroad as the Yorkshire, it is a prick-eared, long-bodied white pig which seems to have arisen from crosses of various breeds onto the large local race. Another possible origin might be the Lincolnshire breed, whose white, long body and prick ears were mentioned by Youatt (1847) and which reached notable perfection in Yorkshire. The crosses probably involved pigs from Leicestershire but the precise origins of these improving breeds are not known. Chinese, Neapolitan, Berkshire and Improved Essex were probably all involved.

The name firmly associated with the emergence of the Large White as a breed of national importance is that of Joseph Tuley, whose exhibit at the 1851 Royal Show at Windsor made a very favourable impression. His reputation was already established as he is mentioned in Richardson's book of 1846. At an early stage it was recognized that the breed really consisted of several strains and some of these were given specific names, such as Coleshill, Improved Manchester, Suffolk, and Middlesex or Prince Albert or Windsor breeds. This confusion contributed to the foundation of the National Pig Breeders' Association in 1884 which was established to meet the criticism from American buyers that the new white breeds were not true to type. The breeds under the aegis of the Association from the outset were the Large, Middle, and Small Whites or Yorkshires, and the Small Black. The last two mentioned soon ceased to be recorded.

Middle Whites were developed in step with Large Whites, breeders of one often keeping the other too. Always recognized as an early-maturing pork breed, these pigs show the strongest influence of Chinese blood, in the deeply dished face and smaller size than the Large White. This illustration is from Sanders Spencer's book *Pigs. Breeds and management* (1897). Spencer, an extremely influential breeder of pigs, was called the 'Bakewell of pigs' by Watson & Hobbs (1951). The Middle White breed provides a vivid example of the value of conservation work done and supported by the Rare Breeds Survival Trust. There is now a thriving premium meat trade based on the Middle White and other rare breeds. The Heal Farm enterprise of Mrs Anne Petch in Devon is the best known.

Below: Sir Stanley Spencer RA (1891–1959) painted these Large White pigs at Rickett's Farm, Cookham Dene, in 1938. While British pig farming was kept subsidiary to other forms of agriculture, marked progress in pig husbandry was not possible. With the development of intensive units, efficiency was substantially increased. Nowadays public disquiet over the welfare of pigs may lead to changes back to less intensive systems, although Rickett's Farm is hardly likely to be a model for the future.

Like the other pedigree pigs the Middle White attained great popularity in the 1920s but since 1945 its popularity crashed and now it is Britain's rarest breed of pig. However, it has made its mark, and as the extreme pork breed it is of the greatest conservation importance as it represents a genetic resource for the future.

These two breeds have, like the Berkshire, had a tremendous influence on the world's pig population. Davidson (1966) pointed out how the Berkshire 'travelled west to the land of lard hogs' while the Large White has tended to migrate south and east into Europe, being particularly heavily used in France, Russia, and Germany. In fact the value of the Large White as a crossing breed has spread even as far as the Far East, which two hundred years ago made such a marked contribution to the pigs of the western world. Indeed, in October 1986 a record

Suffolk and Bedfordshire pigs painted by J.W. Giles. The Prince Consort, hatted, carries a pitch of straw on a fork. Windsor Castle is in the background. These three twenty-five week old pigs gained the second prize of £5 at the Smithfield Show in 1844, having been fed on milk, barley-meal, and pea-meal.

3900 guineas was paid for a Large White boar by a Japanese buyer. While the Middle White was, in the early decades of the twentieth century, one of Britain's most important breeds both numerically and in terms of exports, particularly to the Far East, the Large White had a more lasting influence. Many individual strains exist in the breed and a spectacular example of how the right variety can meet a real commercial need was provided by the Large Whites sold by Sanders Spencer to Denmark in 1899. The local breed, the Danish Landrace, was at that time a prolific but coarse-bodied animal. When mated with the Large White boar it gave excellent carcasses which by 1924 represented 50 per cent of British bacon imports. This new industry was founded on clear-sighted scientific and commercial principles and on the effectiveness of the Large White—Landrace cross.

Forerunners of the Large White were the massive pigs of Yorkshire and over much of the world the Large White is usually known as a Yorkshire. This painting, by an unknown artist, was engraved by R. Pollard and published in 1809. The caption is as follows:

'The Yorkshire hog, to Coln. Thomas Charles Beaumont Esqr. M.P. This specimen of an improved breed of this useful animal was bred by Benjn. Rowley of Red House near Doncaster & fed by Josh. Hudson on the estate of Col. Beaumont of Bretton Hall to whom this print is respectfully inscribed by his obedt. humble servant Joseph Hudson. This stupendious creature for height & length far exceeds any of this species ever yet seen measuring 9′ 10′ long 8′ round the body stands 12½ hands high 4 years old & weighs 1344 lbs. (or 160 stone 8 lb to the stone or 96 st 14 lb to the stone) & would feed to a much greater weight were he not raised up so often to exhibit his stature. He has been view'd by the Agricultural Society & the best judges with astonishment & excited the public curiosity so much that the proprietor for admittion [sic] money to see him in 3 years has received near 3000 pounds!!!'

215

Tamworth

The early Berkshire, illustrated on p. 207, looks quite like a present-day Tamworth, and the simplest explanation of this would be that the Tamworth is an unimproved Berkshire. Whether the red colouring is an ancestral feature, common to both breeds, or is the result of imports, is unlikely ever to be determined. There are accounts of red pigs being brought into the Midlands, one type being the 'Axford pig', a red or red-black pig from Barbados, apparently used around 1750 near Marlborough; also a wild type pig reportedly brought from India to Tamworth around 1814. Sir Robert Peel, who lived near Tamworth, may also have brought pigs there from Ireland around 1812.

The Tamworth had a definite identity of its own by 1860. Sidney (1871) described it as a red or red-black breed falling in popularity because it would not fatten. This late maturity, and its long snout, with no hint of a dished face, indicates a lack of Chinese or Neapolitan influence. Later in the nineteenth century, however, the Large White was used to improve the Tamworth's commercial qualities. By 1909 the breed had been stabilized in form and colour, but it had been ousted by the Large White from the competition to find favour with pig breeders seeking leanness of carcass.

'A breed that should be encouraged' according to Layley & Malden (1935) because of its merits as a bacon pig, but this was not to be, and the Tamworth is currently one of Britain's rarest pigs. This illustration is by Tunnicliffe.

The Tamworth was exported extensively to Canada, the USA, Australia, and New Zealand but has for long been a minority and latterly a rare breed in Britain. In 1985 only seventeen boars were born (compared with 165 registered in 1949). In 1979 the Rare Breeds Survival Trust re-imported three Tamworth boars from Australia. It seems likely that the Tamworth is destined to remain a rare breed, favoured for its appearance and dark skin which does not sunburn, but any move towards outdoor pig-keeping is more likely to favour the numerically stronger British Saddleback.

Tamworth sow, golden in the late afternoon sun.

Other white breeds

The Gloucester Old Spot could easily have gone the same way as the Lincolnshire Curly Coat. By 1961 the latter breed numbered only twenty-seven registered pigs including one boar; there were only five active breeders, and the breed society closed down. By 1972 the breed was extinct, though its image lives on in a few photographs and in this illustration by Tunnicliffe. We can infer the reasons for its demise from Layley & Malden's description in 1935 of the market to which it most appealed; 'sturdy and hardy workers in its county (who) have not altogether lost their appetite for well-cured and matured bacon fortified by a thick firm layer of fat such as the more southern townsman dare not face'.

In the north of Britain, by the time the National Pig Breeders' Association was formed in 1884, most of the local varieties had been merged into the Large White. A few races remained and their breeders decided their future was as independent breeds, and societies were formed for the Ulster, the Lincolnshire Curly Coat, and the Cumberland in 1907, 1908, and 1915 respectively. Further south, societies for the Gloucester Old Spot, the Welsh, and the Long White Lop became established in 1914, 1918, and 1921. The Welsh was known for a while around 1920 as the Old Glamorgan. Other races retained the loyalty of local breeders even though formal breed societies were either not established, or did not last for long, and these included the Yorkshire Blue and White, the Oxford Sandy and Black, and the Dorset Gold Tip. These minor varieties were all still extant in 1949 when the British (previously Long White) Lop and Lincolnshire Curly Coat were the most numerous with 85 and 82 boars registered respectively. The Cumberland, Oxford Sandy and Black, and Dorset Gold Tip were the rarest with five, seven, and two boars licensed respectively. The four other breeds named here each numbered around thirty licensed boars.

By 1973 all were considered extinct except the Welsh with 671 boars licensed, the Gloucester Old Spot, and the British Lop. Whether the Oxford Sandy and Black escaped extinction is debatable; today its adherents claim present-day pigs of this type can trace their ancestry to the early years of this century.

The photographs in two of the standard works of sixty years ago – the Ministry of Agriculture's *British Breeds of Live Stock* (1920) and Sanders Spencer's *Pigs. Breeds and management* (1897) depict Gloucester Old Spot pigs with a great deal of black spotting. The 1933 Pigs Marketing Scheme discriminated against coloured pigs and the breeders reduced the number of spots aiming at only one large spot on each side. This painting by James Lynch was made in 1982.

This photograph and the skeleton of a Gloucester Old Spot pig were donated to the British Museum (Natural History) in 1921. Though outwardly a fine boar, Gilslake Major's limb bones were seriously diseased even though his age at death was only three years.

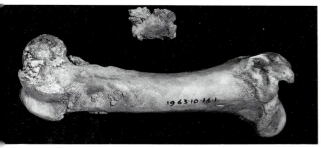

The Welsh and the British Lop were both influenced by the Landrace. Landrace pigs are a type which exists as different breeds in Sweden, Norway, Germany, Finland, the Netherlands, Belgium, Britain, and the USA. They are long in the body, with lop ears, and are white in colour. They were developed following the aims and breeding methods defined by the Danish pig improvers at the start of this century, to produce pigs suitable for the Wiltshire bacon process. They first came to Britain in 1953. Now, Welsh and Landrace pigs are very similar, and the present British pig industry is almost completely dependent on these breeds, together with the Large White and the cross-breds and hybrids between them.

The British Lop has not attained the same commercial significance as the Welsh, remaining a local breed, particularly well represented in Cornwall but with some notable herds elsewhere. In the last few years only about sixteen herds have been registering these pigs (for the other rare breeds the 1985 figures are as follows; Berkshire 22, Tamworth 23, Middle White 15, Large Black 20, Gloucester Old Spot 41, and British Saddleback 20).

Today the Gloucester Old Spot is numerically the healthiest rare pig breed. There are around 600 sows with 90 breeders, although only a proportion of these (41 in 1985) register their stock. Before 1914 the breed was hardly known outside its area and classes were first held at the Royal Show in 1919 at Cardiff. In 1909 the Royal Show had been at Gloucester and it is interesting that no mention was made of what is presumably to be considered as the county's indigenous breed.

The ways in which the Gloucester Old Spot arose from the ancestral stock are not documented, and one suggestion by Wiseman (1986) has been that it owes something to the Large White. Its breed society was formed in 1920 and the new breed prospered in the 1920s as a dual-purpose pig although in the Ministry of Agriculture's book of 1920 its use as a bacon pig was stressed. Its humble origins are implied by the alternative name it had then of 'Cottage pig'.

The breed today has been dominated by stock of the Ribbesford herd of G.H. Styles & Son. Between 1979 and 1986 this herd supplied 91 boars to other Gloucester Old Spot breeders which sired registered stock in their new herds. In this period 140 were active and 85 of them used Ribbesford boars. This is unlike the other rare breeds of pig where there is not such a strongly defined hierarchy of herds. Since the early 1970s the emphasis in the Ribbesford herd was on indoor pigs and the present Gloucester Old Spot is, on the whole, kept under such conditions. The tests carried out on the pure breed in comparison with the Large White, at the Meat Research Institute, showed the latter to be superior in all respects as a pure-bred meat producer. This is not, however, as bad a result as it may appear from the point of view of the Gloucester Old Spot because most pigs in Britain today are cross-bred and it is the quality of the cross-bred pig that is important. Most Gloucester Old Spot herds now use Large White or Landrace boars to sire the slaughter generation.

In the early nineteenth century British white pigs, exported to Chester County, Pennsylvania, in the USA, formed the basis of the Chester White breed. These pigs seem to

This Welsh sow, Ashdonian Imposing 76th, was bred by Mr Ketteridge of Saffron Walden, Essex, in 1980. This embroidery, by Helen Stevens, was commissioned by him. This pig was champion in the East of England heat of the Barclays Bank Pig of the Year competition in 1981.

have particularly good maternal qualities with high prolificacy and weaning success, and an ability to do well without elaborate housing. They were re-imported to Britain by Mr Geoffrey Cloke in 1978. It is at least possible that the genetic make-up of this breed includes contributions from the Lincolnshire Curly Coat, the Cumberland, and the forerunners of the Large White.

ASHDONIAN IMPOSING 76TH

Large Black

In the mid nineteenth century pigs judged at the Royal Show were classified as 'large breed' or 'small breed'. The latter seems to have arisen from crosses of Chinese or Neapolitan pigs on the native stock and these small pigs were favourites of many of the Victorian fancy breeders. Most of these small breeds failed to satisfy commercial needs and died out before 1914. One reason, according to Layley & Malden (1935), was the total dependence of these pigs on high feeding, and their 'extreme delicacy'. Among these casualties were the small black pigs of the south and east of the country, excepting the Berkshire. Some authorities rejoiced, notably Layley & Malden who felt the Chinese importations had done nothing but harm.

From the confusion of strains, breeds, and races there had emerged the Large Black breed, which certainly did satisfy a need and by the time its herd book was established in 1898 it was physically and numerically one of the largest breeds of pig in Britain. Two strains, one from Devon and Cornwall, the other from Suffolk and Essex, were amalgamated to create the breed, which was defined by Hammond in the 1920s as intermediate between pure pork and bacon types (like the Gloucester Old Spot, British Lop and British Saddleback). Intermediate pigs have often found favour as the farmer has the choice of sending animals for slaughter for pork when prices are good, or of keeping them a few more weeks if the bacon trade is strong.

The Large Black has always had lop ears and a face which is not particularly dished. It was long felt that the West Country and Eastern Counties strains were different, the former being rather larger and coarser. The breed proved very suitable for outdoor and small scale pig husbandry as well as satisfying larger farms particularly when cross-bred. The cross with a Middle White gave, in the 1920s, medium sized pork or bacon pigs 'suitable for the London trade', while matings with the Large White gave a larger type of pig, 'in demand in the Midlands and northern England'.

High prices were paid for Large Blacks at this time, with large numbers being exported. One sow, Bixley Bountiful, sold for 600 guineas in 1920. During the 1920s, there were 2000 members of the Large Black Pig Society.

The Large Black is the breed which suffered most markedly from the changes that took place after 1945 in British pig farming. In 1949, 926 boars of this breed were registered; in 1985, only thirty-one. In 1966 swine fever was eradicated from Britain allowing the development of large scale breeding and fattening units. The lean white breeds were far more competitive, but the Large Black has a strong reputation of being an outdoor breed and as having particularly good maternal qualities. In the memorable words of a Hungarian delegation who discussed Large Blacks with the Meat and Livestock Commission: 'In our country the housing is bad, the food is bad, the climate is bad and all other breeds die'.

Large Black pigs at Litlington, Hertfordshire. At this traditional outdoor pig farm, Large Black sows are mated with Large White boars to produce crossbred litters for slaughter. This cross exploits the mothering qualities of the former with the good carcass qualities of the latter. The boar is being shown here by Charlie Ruddock.

18   Heavy draught horses

Progressive agriculturalists of the period 1770 to 1830 disapproved of farm horses, favouring oxen as requiring less land and providing marketable meat. They failed to convince the farmers partly, at least, because the enormous demand from the cities for dray horses meant that a market was assured for all the horses that were bred. In addition, farmers were able to benefit from a supply of horses of all qualities and at all prices when they were of no further use in the cities. Whitbread the brewers owned this gelding, painted by Garrard in 1792.

The heavy horses of Britain have probably changed less over the last 200 years than any of the other farm animals. Youatt in 1843 described three breeds, the Suffolk Punch, the Clydesdale, and the Heavy Black Horse, now known as the Shire. These horses are still flourishing in Britain, as is the Percheron, a grey heavy horse introduced from France at the end of World War I.

The old Suffolk Punch was said by Youatt to have nearly died out in 1831. The new breed was described as having many of the qualities of its predecessors including the same sorrel (chestnut) colour, but it was taller with higher and finer shoulders as a result of crossing with heavy horses from Yorkshire. All the breeds of heavy horse were crossed with imports from Flanders and the Low Countries (today Belgium and the Netherlands) at the end of the eighteenth century and their size and strength was thereby much increased.

The Shire horse was known as the Old English Black horse until the middle of the nineteenth century, although the modern type was established in Leicestershire before 1790, partly as a result of breeding improvements instigated by Robert Bakewell. The Clydesdale, according to Youatt (1831), owes its origin to one of the Dukes of Hamilton who crossed Lanark mares with stallions that he brought from Flanders. Clydesdales and Shires are the heaviest and strongest horses in the world, reaching a height of 18 hands (183 cm) and a weight of one tonne.

Shire

In contrast to the well-established Suffolk breed, the work horses of the rest of England two hundred years ago did not conform to any type, except that black was a common colour and large size a feature. Robert Bakewell was one of several breeders whose development of the Black Horse of Leicestershire contributed to the new, efficient Midlands type of Shire horse. Imports of big strong horses from the Low Countries brought about much of this improvement and it was the race with the easiest access to these stocks, the fenland type, which was early on noted for its great size. By 1800, using a combination of Bakewell's methods and imported stock, breeders had upgraded the horse of the Middle Ages to a competent work animal, which was, however, not named the Shire horse until rather late in the nineteenth century.

Only a tiny minority of stallions were left entire, as geldings were in great demand for work, and the practice of travelling stallions developed. The selection of males for this purpose led to the recording, at first haphazard, of sire and dam; Honest Tim, foaled in 1800 (bred by William Wiseman of Fleet, Lincolnshire) was the most influential Shire stallion.

Importance was not generally attached to selection and breeding of Shire horses for most of the nineteenth century. For a start, it was a better business proposition to castrate a promising foal and rear him as a gelding than as a stallion for stud. In times of prosperity demand for work horses (at first from farms, then from the hauliers and traders of towns and cities) was considerable and uncritical, while in times of depression there was no money to spare for selective breeding. The Shire suffered, too, from effective competition from the Clydesdale and

'The Old English Black Horse. Stallion, by Old Blacklegs, from a mare of the Dishley Breed; bred by Mr Broomes, at Ormiston, Derby.'
This black stallion, painted by Shiels, was presumably descended from Bakewell's stock of black Leicester horses. At this time, the pedigrees of champion cattle were of much greater interest than those of working cart horses.

the Suffolk. Co-ordinated development of the breed only really began in 1878, when the Earl of Ellesmere was foremost among the enthusiasts who founded the English Cart-Horse Society of Great Britain and Ireland (renamed the Shire Horse Society in 1884). Volume 1 of the Stud Book recorded 2365 stallions. The Royal Show of the following year for the first time offered prizes for horses which were comparable to those offered for cattle.

Pedigree breeding, annual cart horse parades and an enthusiastic breed society all contributed to the progress of the breed. Demand from the United States, where the home-bred horses were far too light for the haulage tasks arising in the new cities, peaked in 1910 when 504 stallions were exported, but began to fall because English show-ring standards, which stipulated heavily feathered feet, were encouraging a type that was unsuited to American conditions. During World War I, in which half a million horses died, as many Shires as possible were bred, but their progeny were not ready for work until 1921. By then they had to compete with ex-Army horses and with tractors and motor lorries; competitions which they were doomed to lose.

Working brewery horses are very much part of the urban heritage. Young & Co. of the Ram Brewery, Wandsworth, have twenty Shire horses. This painting by Nina Colmore (1889–1973) shows the dray in Windsor Great Park in 1953.

It was the town horse that was eclipsed first, as the advantages of the tractor took time and technical innovation to develop. Even so, in the late 1930s there were still about 40 000 draught horses in Greater London as the expansion of the economy and increased demand for transport had compensated for the inroads made by motor vehicles. On the farms in England and Wales between 1922 and 1936 the numbers of horses had fallen by 35 per cent. There were two causes; a 19 per cent reduction in arable area and the fact that each tractor replaced between two and four horses. World War II brought home, finally, the reality that the three acres of land (half grass, half oats) needed to sustain one heavy horse would be better used for human food. Half a million horses were at work in 1947; they were slaughtered at the rate of 100 000 a year. The rate only slowed when there were few left to slaughter.

By 1961 the draught Shire horse was effectively represented by a few brewers' drays, and entries at the Society's Spring Show were at their all time low in that year. In 1963 only 49 new pedigree mares were registered. Since then matters have improved greatly for the breed and for the Shire Horse Society; in 1985 there were 86 registrations of females, and public interest in the breed continues to grow year by year.

Haymakers was painted in 1785 by George Stubbs (1724–1806). It entered the Tate Gallery with its companion piece *Reapers* in 1977. In Milner's words (1983), 'Stubbs's well-scrubbed Sunday-best peasantry scarcely reflect the hot, hurried, dusty and dirty ardour of a real harvest', but this is probably one of the best-known and best-loved English animal paintings.

Clydesdale

James Howe painted his *Scotch Farm Horse* in 1830. Perhaps rather lighter in build than Shiels' 'Old English Black Horse', the early Clydesdale has none the less the same general conformation.

The Clydesdale seems to have originated from a cross of Flemish or English heavy horses on the local stock of the upper Clyde Valley, in Lanarkshire. In many ways its development paralleled that of the English Shire. Both breeds had a history of infusion of Flemish stock; noteworthy sires came into prominence at an early stage; both breeds had their place in a horse-drawn economy where farm rearing and town sale of stock were integrated. Both breeds sold for fantastic prices in the early twentieth century, and both collapsed numerically as motor vehicles took over, to revive during the 1970s.

Indeed, crossing between Clydesdale and Shires has led to a general acceptance that the two are, for practical purposes, different types of the same breed. In both, blood-typing is a prerequisite for the registration of a foal (in Shires since 1981, in Clydesdales since 1984) so the means are available for the breeds to be kept genetically separate if this is desired.

The Stud Book was first published in 1878 and the early records show that, in sharp contrast to many other breeds of livestock where inbreeding is characteristic of the early stages of breed development (including the Suffolk horse), there was very little inbreeding to any of the early founders of this breed. Nevertheless, Clydesdales foaled between 1885 and 1890 were rather inbred, due to a desire to concentrate the blood of three famous stallions, Prince of Wales (bred by Lawrence Drew of Merryton, foaled 1866, died 1888); Darnley (foaled 1872) and Baron's Pride, but the stud book records show that the inbreeding was not intensive, with only one parent-offspring mating recorded. Even so, by 1925 it appears that all registered Clydesdale foals were inbred in some measure to Darnley. Indeed, 'Darnley for mares' became a truism and many outstanding animals came from matings of Prince of Wales to Darnley's mares.

Clydesdales tend to have less feather on the foot than Shires, neater legs and probably a faster action. Horsecollars on Scottish and north-east English farm horses (which are usually Clydesdales) are traditionally peaked as shown here. Shire blood was introduced into the Clydesdale by Lawrence Drew of Merryton between 1870 and 1884. Since 1939 Clydesdale mares have been widely used with Shire stallions in England. This model was made by Judy Boyt in 1983.

It was in north-east England that competition between Shires and Clydesdales was most intense. Up till the 1860s the Shire was particularly popular and after this period was still used occasionally. This Clydesdale stallion, Bonnie Buchlyvie, was bred in the district's most famous stud, that of Lord Londonderry at Seaham Harbour. The manager, Robert Brydon, had a great influence on horse breeding. Bonnie Buchlyvie was sold in 1916 for 5000 guineas.

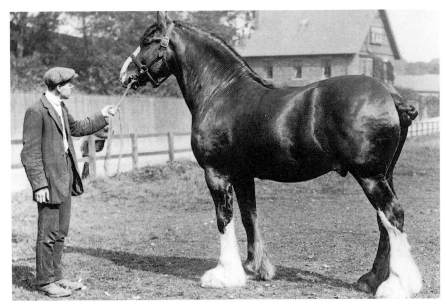

In Peeblesshire and the other Scottish border counties the rearing of horses was a major pursuit; foals were bought at fairs, especially at Carnwath and Lanark, broken-in at Renfrew and Ayr, and resold at Rutherglen and Glasgow for use in eastern Scotland and north-east England. In this painting by Howe (1829) (left) of one of the minor Peeblesshire fairs, most ponies and horses are full grown and ready for work. Grey and chestnut fell out of favour in Clydesdales soon after this picture was painted; bay and brown are now the commonest colours. However, breeding for fancy points was never a preoccupation of Clydesdale breeders, who have claimed to concentrate on the elimination of defects rather than on the development of particular features.

From the 1850s Clydesdales were exported very widely and, just like the Shire, they were used more and more for road haulage. The breed society was founded in 1877. Clydesdales waxed and waned in numbers, almost vanishing in the 1960s, but today the breed is growing stronger. As with the Shire, the danger today is not, as in the early years, that particularly good colts will be castrated rather than retained for breeding; the concern is that nowadays the best breeding stock is being exported, mainly to the USA.

There are today over 1300 members of the Society, 250 of whom breed Clydesdales, with 50 to 60 active stallions and over 300 brood mares, and many more abroad. Probably the most famous of today's Clydesdales are the Anheuser-Busch Budweiser Clydesdales, the Milwaukee 8-horse brewery dray whose owners regularly come to Scotland to buy replacements.

Right: Howe clearly took great pleasure in depicting the action of horses: *Trace Horses on the Turn* was drawn around 1830.

231

Suffolk

The Suffolk has the longest written history of any breed of horse apart from the Thorough-bred. In the eighteenth century it was called the Suffolk Sorrel. Rather smaller than the heavy horse of the Midland counties (that later became the Shire) the Suffolk Sorrel was well suited to general farm work and road use.

The early history of the breed was written by Herman Biddell and published in 1880, as part of Volume 1 of the Stud Book. Biddell traced all Suffolk horses back to a stallion foaled near Woodbridge in 1768, which was advertised thus in a local newspaper in 1773:

> light chestnut horse full $15\frac{1}{2}$ hands [158 cm], five years old, to set good stock for
> coach or road, which said horse is the property of Thomas Crisp of Ufford.

Biddell recorded much information, including heights, and from these we see that the breed has changed very little in the last hundred years. Mares stand at about 16.2 hands (168 cm), stallions 17.0 hands (173 cm). This is rather smaller than the Shire and the difference is in the length of leg. The dumpy appearance is the reason for the nickname 'Suffolk Punch'.

The chestnut colour, or chesnut as the Suffolk breeders spell it, is characteristic of the breed along with the massive, short legs on which only minimal feather is allowed. The only white is on the face. The strong body and neat legs and feet make the Suffolk well suited as a plough horse.

'The Suffolk Punch. Stallion, rising 7, the property of Mr Denny Egmoor, Norfolk.' Low reported that crossing had taken place and it was now 'somewhat difficult to obtain the Suffolk Punch in a state of purity'. None the less the breed today shows a remarkable degree of uniformity of type. It is undoubtedly highly inbred, but the genetic variability that does remain is being conserved by the widespread use of as many stallions as possible. In 1981 the forty-two foals born were sired by fifteen stallions.

Another characteristic has been the strong regional association with the eastern counties; and around the keeping and management of Suffolk horses there grew up a complete culture, whose mystique and romance have been eloquently described by George Ewart Evans (1960). The Suffolk was also firmly associated with large arable farms on relatively light land, and with being kept in horse yards rather than in stables.

In the boom years for heavy horses, before 1914, about a third of all the Suffolks registered were exported, and like the other heavy horses, a great many were sold for town use. Suffolks tended to be used as 'vanners', harnessed singly or in pairs for haulage of light goods. The Shire was more favoured as a dray horse and this is why the brewery turn-outs today are almost exclusively Shires.

Pure-bred and cross-bred mares and foals near Ipswich; exhibited by Sidney Seymour Lucas in 1938.

Farm horses numbered about one million in England and Wales at their peak in 1910, that is about 84.3 horses per 1000 hectares of crops and grass. The highest densities of farm horses were in the eastern counties, with south Lincolnshire and the Isle of Ely at the top of the list. In 1960 there were only 60 000 farm horses at work and even though these two areas still had quite large numbers of horses (they were very useful for potato and sugar-beet cultivation) the pure-bred Suffolk was very hard hit and in 1962 only twelve foals were registered.

The Suffolk has survived, but it is not yet safe as there are only about 24 breeders, with, in 1985, 118 brood mares and 32 stallions, and 24 fillies registered annually. It has not yet had the same resurgence as the Shire horse.

Sudbourne Premier, in bronze, by Herbert Haseltine (1877–1962). The sculptor had made models of twenty British champion animals in the period 1922 to 1934. The original version of this piece (in Burgundy stone) was purchased by the Tate Gallery but was smashed in transit so the Trustees commissioned this replica in bronze in 1932. The stallion was foaled in 1919 and was First and Champion at the Royal Shows of 1921 and 1922.

Percheron

Percheron, by Tunnicliffe
(1940).

Of the twenty or more breeds of cattle, horses, sheep, and pigs whose merits are widely acknowledged throughout the world only three, the Friesian cow, the Percheron horse, and the Merino sheep, originated elsewhere than in our small islands.   (Watson & Hobbes, 1951).

The Percheron represents a combination of Arab blood with that of the Flemish type, under the supervision of the French Government stud at La Pin in the north-west of France. The result was a grey or black heavy draught horse with a minimum of white, and as little hair as possible on the limbs. Breeders have aimed to combine heavy musculature (stallions weigh up to a tonne, mares about ten per cent less) with style and activity. World-wide the breed has been extremely successful but it was only in 1918 that it was first introduced to Britain, 36 stallions and 321 mares being imported. The breed society was founded in the same year. Like the other farm horses the Percheron declined in Britain from the end of World War II; however today British Percherons have shared in the revival of the heavy horse and are being exported widely.

Pony on Dartmoor.

19   Light horses, ponies, and donkeys

Apart from the mountain and moorland ponies of Britain, the evolution of the riding horses has followed a different course from that of other livestock. This is because, being used for long distance travel, the horse has not been restricted to the different regions of the land and so has not been influenced by the processes of natural selection that were of importance, for example, in the breeding of sheep. Moreover, improvement in the horse began much earlier than in other farm animals.

In a history of farm livestock it is difficult to separate the horse used on the farm from the hunter, the pack horse, or the racehorse, because more often than not, one animal was used for all these purposes. The Cleveland Bay, for example, is a breed that was originally a farm animal but later became much valued for breeding carriage horses and hunters.

The mountain and moorland ponies more closely resemble the breeds of cattle and sheep of Britain in that they have evolved in and been restricted to certain regions of the country. These ponies lived half wild and were rounded up once a year or so to be used on farms, sold, branded, or castrated. Over the last hundred years there have been attempts to improve most of the races of these ponies by interbreeding with Arab sires and, much earlier, it is likely that attempts were made to breed for a lighter-built, less stocky appearance. There are two legends that recur about the ponies; one is that they are 'Celtic' in origin. This term appears to have a variety of meanings from ancient British to Spanish to Norwegian. The second legend claims that ponies on the western margins of the British Isles, particularly the Connemara from Ireland and the Galloway from the Scottish borders, are descended from Spanish horses rescued from the Armada in 1588. Both legends may have an essence of truth in that the half wild stock of native ponies is probably of ancient origin and has received little outside blood until relatively recent times. It is also likely that ponies of Arab descent were brought by land-owners from Spain and other Mediterranean countries to improve the local breeds over the last 300 years.

Dartmoor, Exmoor, and New Forest ponies are and always have been riding ponies, though capable of carrying a load when required. They do not really count as farm livestock and the same applies to the Welsh ponies and the Connemara, though it would be a different matter if horsemeat had ever become an important item in the British diet. Today Britain produces about 6000 tonnes of horsemeat a year, most of which goes to the Continent. However, it would be a limited perspective on farm animals which did not recognize the importance to British farming of the half a million or so horses and ponies in Britain, practically all of which are kept for recreational purposes. The membership of the British Horse Society today is

36 500 and has been increasing at the rate of 4 to 6 per cent per annum over the past five years.

All the native British ponies have been used as pack animals and for general farm work, but today, apart from their use as riding ponies, they are protected as part of the large mammal fauna of the few wild places where people can still go to get away from the turmoil of urban life.

Masterful Jack, a fine Cleveland Bay stallion four years old and standing at 16.3 hands (170 cm), was photographed in 1988 at North Medburn Farm, Elstree, Hertfordshire with his owner Mrs R.O. Stevens.

Cleveland Bay

'The Cleveland Bay. Stallion, by Catfos, the sire of Bay Chilton; dam by Mr Ayres' Rainbow, bred by Mr Robertson of Naperton, Holderness.'
By 1840 when Shiels painted this picture, the Cleveland Bay had absorbed the influence of Thoroughbred stallions. The Farmer's Magazine in 1826 described it as 'neither Blood nor Black, that is, a distinct race from the English Blood-Horse, and equally distinct from the Black or Cart breed of the country'. The breed was used as a carriage horse in the early nineteenth century, but was superseded by lighter horses and nearly died out except in Yorkshire. It continued in use particularly on the lighter soils of the Whitby area until about 1900. By then it was famous once again (with its offshoot the Yorkshire Coach Horse) as a carriage horse and heavy hunter.

Masterful Jack, a fine Cleveland Bay stallion four years old and standing at 16.3 hands (170 cm), was photographed in 1988 at North Medburn Farm, Elstree, Hertfordshire with his owner Mrs R.O. Stevens.

The bay horses of Cleveland in Yorkshire were renowned until the early part of this century for every kind of draught work. They were often known as Chapman horses because they were used to carry the loads of the travelling salesmen or chapmen around the countryside. Originally the Cleveland Bay was a strong, broad-backed agricultural and pack horse, until in the eighteenth century the breed was improved by crossing with Thoroughbred stallions descended from the Darley Arabian and the Godolphin Arabian. The progeny were larger and showier than their forbears and in the nineteenth century they became popular carriage horses, a breeding line which later became known as Yorkshire Coach Horses. By the end of the century there were two branches each with its own Stud book, the Cleveland Bay founded in 1884 and the Yorkshire Coach Horse founded in 1886. Yorkshire Coach Horses were bred from pure Cleveland Bay mares crossed with cross-breds and they were extremely popular carriage horses until the motor car brought an end to this form of transport. The Yorkshire Coach Horse Society was closed in 1937; in 1960 the Cleveland Bay was at its lowest ebb and it is thought there were only three pure-bred stallions at stud. Now, the breed has been saved from extinction and is again popular for cross-breeding with Thoroughbreds to produce hunters, show-jumpers, coach and event horses.

A TYPE OF FARM HORSE OF THE "VARDY" OR "BAKEWELL" BREED,

INTRODUCED INTO NORTHUMBERLAND BY THE MESSRS. CULLEY AT THE CLOSE OF THE LAST CENTURY.

The above figure represents the portrait of the grey gelding, bred by the late Mr. Curry, of Brandon, Northumberland, and was the property of Messrs. Howey & Co., the great Carriers from Edinburgh into England. This Gelding exhibits such a form as to constitute, in my estimation, the very perfection of what a farm horse should be. His head (A) is small, bone clean, eyes prominent, muzzle fine, and ears set upon the crown of the head. His neck rises with a fine crest from the trunk (B H to A), and tapers to the head, which is beautifully set on the neck, and seems to be borne by it with ease. His limbs taper gradually from the body, and are broad and flat, indicating strength; the fore arm (I) broad and flat; all excellent points in the leg of a draught horse, giving it strength and action. The back of the fore leg, from the fetlock joint (L) to the body (O), is straight, indicating no weakness in the limb—a failing here causing the knees to knuckle, and rendering the horse unsafe in going down hill. The hind legs (M), as well as the fore ones (K and L), stand directly under the body, forming firm supports under it. The body is beautifully symmetrical. The shoulder slopes backwards from H to B, the withers at B being high and thin. The sloped position of the shoulder affords a proper seat for the collar, and provides the muscles of the shoulder blade (G) with a long lever to enable them to throw the fore legs easily forward, and with such a shoulder a horse cannot stumble. The back, from B to C, is short, no longer than to give room for the saddle. The chest, from B to O, is deep, giving capacity for the lungs to play in, and room for the muscles required in draught.

The top of the quarter, from C to D, is rounded; the flank, from C to N, deep; and the hind quarter, from F to E, long. On looking on the entire side profile of the animal, the body seems made up of two large quarters, joined together by a short thick middle, suggesting the idea of strength and action; and the limbs, neck, and head are so attached to the body as to appear light and graceful. In a well-formed horse, I may remark that the line from the fetlock joint (L) to the elbow joint (O) is equal to that from the joint (O) to the top of the withers (B). In a low-shouldered leggy-horse, the line L O is much longer than the line O B; but in the case of this horse, the body (B O) is rather deeper than the leg (L O) is long, realising the desideratum in a farm horse of a thick middle, and short legs. The line across the ribs, from G to F, is like the back, short, and the ribs are round. He was 16 hands high, or 64 inches; measured from A to B, 35 inches, and B to C, 33 inches; from C to D, 19 inches, being in extreme length, 7 feet 3 inches. Length of the face, 25 inches; breadth of face across the eyes, 10 inches; length of ears, 6½ inches; breadth across the hook bones, 22 inches; girth behind the shoulder, 80 inches; girth of fore arm, 23 inches; girth of bone below the fore knee, 9½ inches (the girth of this bone shows the comparative strength of the fore leg of every horse); girth of the neck at the onset of the head, 32 inches; girth of muzzle, 21 inches; width of counter, 19 inches; and height of top of quarter (C) from the ground, 63 inches. In a draught horse, the use of the collar causes the muscles upon the shoulder to enlarge and the neck to become thin. This horse's name was "Farmer;" his walk was stately, and he could draw 3 tons on level ground, including the weight of the waggon. He was a well known horse on the streets of Edinburgh for some years, and was generally admired. He was 11 years old when his portrait was taken in 1838.

This Plate is the Gift of JOSEPH SNOWBALL, Esq., of Seaton Burn House, Dudley, Northumberland, to the Members of the Northumberland and Coquetdale Agricultural Societies, and is Engraved from a Drawing by John Sheriff, A.R.S.A., and Published in Stephens's "Book of the Farm," from which the above is an extract.

The Vardy Horse was developed in north-east England from crossing horses of Cleveland Bay type with Shire horses, for which a connection with Bakewell's stock was subsequently claimed. It was a small cart horse rarely over 16 hands (163 cm), popular around 1830 to 1850 in Northumberland but too small, compared with Clydesdale or Shire, to be valuable for town use. By 1910 it was extinct.

Overleaf: Every autumn the ponies of the New Forest are rounded up, the new foals are branded, the herd inspected and wormed, and surplus stock is sold off. Lucy Kemp-Welch (1869–1958) painted the 'colt-hunt' or 'drift' in 1897.

English ponies

The Fell pony performed the same role, that of all the work of the hill farm, on the west side of the Pennines, as the Dales Pony did on the east. Taylor Longmire has painted a Fell pony with a selection of Lakeland stock in Troutbeck Valley, from the Kirkstone Pass side. Fell ponies (up to 14 hands; 142 cm) have no white markings and are rather smaller than Dales ponies (up to 14.2 hands; 147 cm) which may have some white about the head and feet. The two breeds are moving apart but even in 1977 a number of ponies were registered in both Stud Books.

Ponies living in a half wild or feral state have probably inhabited the moorlands of England and the New Forest for hundreds and perhaps for thousands of years. They are probably descended from domestic ponies introduced during the late Neolithic period, in about 2000 BC. There must have been constant introductions from outside stock since then, but over the centuries each region has developed its own race of pony as a result of adaptation to the harsh local environment and artificial selection for agricultural and draught purposes. Some out-breeding with the intention of improving the size, strength, and appearance of the ponies has undoubtedly occurred, but it appears to have had little effect on the conformity of the types. There are five main breeds of native English pony; the Exmoor, Dartmoor, New Forest, Dales, and Fell. The Exmoor has the reputation of being the most primitive of these breeds because in

The Dales pony is a 'miniature cart horse of an active kind' according to the Ministry of Agriculture's *British Breeds* (1920). In this photograph, the Dales pony is pulling a sledge laden with peat. Until about 1870 they were heavily used as pack horses by the Pennine lead mines.

its small size (11 to 12.2 hands; 112 to 127 cm), dark colour and mealy muzzle it resembles the Tarpan, the now extinct wild pony of Europe, Russia, and Central Asia.

Dartmoor ponies are also small and dark but they have a more mixed origin than the Exmoor. The New Forest has been the subject of controlled breeding and improvement; during the 1850s stallions were lent by Queen Victoria to the Verderers of the New Forest in order to improve the stock, and since that time a very wide range of sires has been used for cross-breeding. Since 1945 the stock has improved markedly, and a general type has emerged. However, heights can still range from 12.2 to 14.2 hands (127 to 147 cm).

Dales and Fell ponies derive from the northern counties; they are larger than their southern counterparts and have been used for interbreeding with heavier farm and haulage horses. Dales ponies are associated with the eastern side of the Pennines, Fell ponies with the west. These two breeds were more genuinely farm animals than other British pony breeds. They were very similar at the time the Dales Improvement Society was formed in 1916, which aimed to increase size to 15 hands (152 cm). There had already been a good deal of Clydesdale influence on these ponies. The Society felt this had gone far enough and the time had come to breed pure, although individual breeders continued to cross-breed.

Exmoor was a Royal Forest until 1818 and the last Warden (hereditary custodian) was Sir Richard Acland. When the area was disafforested Acland drove four hundred of the ponies onto his own property and these gave rise to the Anchor Herd which is still thriving. This photograph shows them in their native setting.

Great attention was paid by the breeders of Dales ponies to the soundness of the limbs. Today these ponies are mostly jet black, brown, bay, or sometimes grey, but chestnut and broken colours are never seen in pure-bred stock. The advent of lorries and tractors nearly put paid to the Dales pony and a low point was reached in 1955 when only four ponies were registered. The Society was reorganized in 1963 and since then numbers have increased.

The co-existence of free-ranging ponies with other large mammals (cattle, goats, deer) in such nature reserves as the New Forest and the Isle of Rhum (Inner Hebrides) gives zoologists an opportunity to study interaction and competition among these species, and they are therefore of considerable scientific value.

Welsh ponies

Youatt in 1831 described the Welsh pony as 'one of the most beautiful little animals that can be imagined'. The renowned good looks of the Welsh Mountain pony may go as far back as the beginning of the eighteenth century, when a small Thoroughbred named Merlin was reputed to have been turned out on the hills of Denbighshire to breed with the native mares. At the beginning of this century the famous grey stallion Dyoll Starlight added another refinement to the breed. The Welsh Mountain pony today has to stand at less than 13.2 hands (137 cm) but it is probably the most valuable of the native breeds for the production of riding ponies.

A second more stocky horse from Wales was the Welsh Cob which was renowned for its fast trotting gait. In 1902 the first stud book of the Welsh Pony and Cob Society was published and at this time the Cob was a widely used working pony on the hill farms.

The Welsh pony is noted for its excellent action. This is a Section A pony up to 12 hands (122 cm). The Section B pony is up to 13.2 hands (137 cm).

This famous Welsh Cob, Old Douse, was painted by Sawrey Gilpin RA (1733–1807), father-in-law of George Garrard.

Bryn Arth Madonna, a Welsh Cob mare of twenty-four years, at work with her owner Mr David Rees in Cardiganshire. The photograph was taken in the early 1960s.

The celebrated stallion Dyoll Starlight had a tremendous influence on Welsh ponies. He was foaled in 1894 and died at the age of 35 years. His skull is kept in the British Museum (Natural History).

It is thought his dam, Dyoll Moonlight, was descended from the Crawshay Bailey Arab (1850). Certainly he represented a beautiful and extremely successful combination of Arab horse and mountain pony.

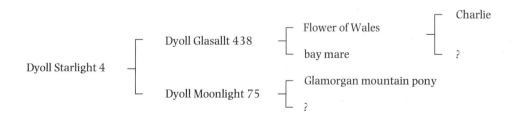

Scottish ponies

Hebridean ponies painted by Shiels: a grey gelding from Mull, a bay mare from Barra and a dun mare from Uist. Much crossing with mainland stock has effaced most local differences among island breeds, but on Eriskay (off South Uist in the Outer Hebrides) a distinctive type has been identified and is now being conserved.

There were three main races of native ponies in Scotland; the Galloway, the Highland, and the Shetland. The Galloway, which was extinct by the end of the nineteenth century, was a large pony that originated in the north of England and southern Scotland, and tradition holds that the breed was derived from horses rescued from the Armada because of their size, length of limb, and elegant appearance. Low, however, claimed in his book of 1845 that the breed was distinguished long before the Armada but it is hard to tell on what evidence. Youatt (1831) stated that the breed was fast degenerating but that relics of it survived on the Isle of Mull. In 1920 the Ministry of Agriculture's book *British Breeds* stated that large Galloway type ponies were still to be found on some of the Inner Hebrides.

Today the Highland pony is considered to be a greatly superior breed but in Youatt's time it was, according to him, far inferior to the Galloway, being large-headed, short-limbed and not pleasant to ride. It was used as a pack pony and for general farm work. At the present day the Highland pony often has a finch-back which probably derives from interbreeding with Norwegian Fjord ponies at some time in the unrecorded past.

Contrary to his view on the Highland pony, Youatt considered that the Shetland pony is 'often exceedingly beautiful, with a small head, good-tempered countenance . . . legs flat and fine, and pretty round feet'. Nowadays the breed's greatest value is as a pet, both in driving and teaching young children to ride, but in the past the Shetland was used for ploughing and general work around the crofts. Robert Brydon, agent for Lord Londonderry whose Seaham Harbour stud of Clydesdales was world famous, bought very many Shetland ponies for colliery work. He noted that the quality of the ponies was degenerating owing to the lack of a proper breeding policy on the islands, and in 1870 he persuaded Lord Londonderry to buy the Shetland island of Bressay as a breeding centre for these ponies. The stud was dispersed in 1899 and by that time had made its mark on the breed; pedigrees of many of today's Shetlands can be traced back to Bressay animals.

The small size of the breed was probably an adaptation to the harsh and barren conditions of the Shetland Islands where there would be very little fodder in winter. The Shetland Pony Stud Book Society was formed in 1890 and it has continued registrations ever since.

Today Highland ponies, like this one painted by Susan Crawford, are widely used in stalking Red deer, the only remnant of their historic use as pack animals. This is a scene on the estate of Sir Donald Cameron of Lochiel at Achnacarry, Inverness. The stag was shot by the artist's husband. Stalking-ponies like this one carry up to 8 stone (114 kg) for long distances over difficult country during the stag and hind shooting seasons which last from the end of July to January. Typically, power-fully-built animals are chosen and trained from the age of three years. Indeed, the foal often follows its dam out to work.

Connemara ponies

The Connemara of Low's day was being crossed extensively with exotic, Barb or Arab stallions with the approval of expert opinion. Shiels' painting clearly shows such an influence. By the end of the nineteenth century further fresh blood was felt necessary and Welsh stallions were imported. This gelding from Galway was 13½ hands (137 cm) high.

Tradition holds that the Connemara pony is descended not only from Celtic stock but also from Spanish horses saved from the Armada. Low in 1845 wrote that the ponies from the west of Ireland were almost unknown in England. They stood at 12 to 14 hands (122 to 142 cm) and were of a prevailing chestnut colour. They appear to have lived wild like other moorland ponies and were caught by being driven into the bogs.

In 1900 J. Cossar Ewart was commissioned by the Department of Agriculture and Technical Instruction to make a survey of the Connemara ponies. Ewart travelled to the west of Ireland and found a breed of ponies that had traditionally been used for every kind of work on the mountain farms and was exceptionally strong and hardy.

Today the colours of the ponies are mixed, following considerable cross-breeding at the end of the nineteenth century. In 1923 the Connemara Pony Breeders Association was established to conserve and improve the breed, and since then the ponies have become popular riding ponies and have been exported to many countries.

Donkey

In 1845 Low wrote, 'In the British Islands, asses are in great numbers, chiefly used by the poorer classes'. He continues with the interesting comment that, 'Great numbers of she-asses are kept about London and the larger towns, for the purpose of supplying a mild, salutary, and nutritive liquid to the infirm.' However, it is only in Ireland that the donkey attained some supremacy. In 1893 there were estimated to be over 200 000 donkeys employed in Irish farming. Dent (1972) maintains that this was because during the eighteenth century Ireland was denuded of horses by the demands of the Napoleonic Wars. In England the donkey was of very low status and in London in the 1890s the average beast was worth from £2 to £5, while a good pair of carriage horses was worth about £400.

The domestic donkey is descended from the wild ass of the Arabian and Saharan deserts. For this reason the British Isles are too northern and too humid for the donkey to adapt to very successfully as a farm animal, and mule-breeding has never become a tradition for producing pack animals. Nevertheless donkeys have played a part in haulage and as servants of tradesmen particularly in the dairy business.

When Tegetmeier & Sutherland wrote their book *Horses, Asses, Zebras, Mules and Mule Breeding* in 1895 they pointed out how no book on the mule had been published although there were 4000 titles on horses and their uses, half of which had been printed in Britain. With 'no ancestry, and no hope of posterity' the mule excited the interest of few agriculturalists. Even so, some enthusiasts took up mule-breeding and working, and notable among these

In Poitou, France, there has long been an industry for producing very large mules from 'giant' Poitou donkeys. Poitou mules have occasionally been imported into Britain. One such, Brunette, stood at 16.1 hands (165 cm) and was very successful at the agricultural shows between 1875 and 1881.

The donkey is, in Wilson's words (1916), 'the plaything of a richer man's children or the slave of some costermonger . . .' This is well illustrated by Sir Thomas Lawrence's painting of the Pallisson children and their pet (right) which is in sharp contrast to Mrs A.M. Hall's snapshot (overleaf) of a lady in the west of Ireland with her helper in 1938.

253

Co. Fermanagh, Ireland, 1924. This donkey is seen carrying peat in wicker paniers. Today, worldwide, there are nearly twice as many working donkeys (39 million) as there are tractors (21 million) and the donkey is inseparable in many countries from small scale husbandry.

This intriguing picture, by an unknown artist, of a mule train travelling from Merthyr to Cardiff in South Wales, is dated 1794. Most appear to be carrying coal, and others carry either wool or cloth. These small mules, 12 or 13 hands (122 to 132 cm), would be able to carry a payload of about 200 lb (91 kg).

in the late nineteenth century was the Duke of Beaufort who used teams of large mules for farm and estate work at Badminton, and Mr A.J. Scott who bred large mules from English cart mares and foreign jacks and used them on his estate near Alton in Hampshire. He also exported several jacks and jennies. In Ireland, too, mules had been bred for many years and Andalusian jacks were imported in the early 1890s by the government. For a few years during the nineteenth century donkey shows were held in the Agricultural Hall, Islington, but interest waned until 1967 when the Donkey Show Society was founded, later changing its name to the Donkey Breed Society. The revival of the donkey, mainly as a pet, has led to the breeding of particoloured animals, and their numbers in Britain are still on the increase.

Bibliography

There is a great wealth of literature on the history of farm animals since the eighteenth century and only a small proportion can be given here. References that are cited in the text are listed below in alphabetical order. These are followed by a selection of references for general reading under the relevant subject headings.

CITED REFERENCES

Alderson, G. L. H. 1978. *The chance to survive: rare breeds in a changing world.* London: Cameron & Tayleur.

Bailey, J. & Culley, G. 1794. *General View of the Agriculture of the County of Northumberland with observations on the means of its improvement.* London.

Baudement, H. E. 1861. *Les races bovines au Concours Universel Agricole de Paris en 1856. Etudes Zootechniques.* 2 vols. Paris: Imprimerie Imperiale.

Bewick, T. 1790. *A general history of quadrupeds.* Newcastle upon Tyne: Edward Walker.

Black, J. 1784. The use of the Roman ox-yoke. *Transactions of the Society for the Encouragement of Arts Manufactures and Commerce* **2**: 81–93.

Blaxter, K. 1986. *People, food and resources.* Cambridge: Cambridge University Press.

Bowie, G. 1987. New sheep for old – changes in sheep farming in Hampshire, 1792–1879. *Agricultural History Review* **35**: 15–24.

Boxall, J. P. 1972. The Sussex breed of cattle in the nineteenth century. *Agricultural History Review* **20**: 17–29.

Bradley, E. *Quoted by* Low, D. 1842. In: *The Breeds of the Domestic Animals of the British Isles.* Vol. I. *The Horse, and the Ox.* London: Longman, Orme, Brown, Green & Longmans.

Chaplin, C. *Quoted by* Perkins, J. A. 1977. In: *Sheep farming in eighteenth and nineteenth century Lincolnshire.* Grimsby: Society for Lincolnshire History and Archaeology Occasional Paper No. 4.

Clark, J. 1794. *General view of the agriculture of the county of Radnor, with observations on the means of its improvement.* London.

Cobbett, W. 1830. *Rural Rides in the counties of Surrey, Kent, Sussex, Hampshire, Wiltshire, Gloucestershire, Herefordshire, Worcestershire, Somersetshire, Oxfordshire, Berkshire, Essex, Suffolk, Norfolk, and Hertfordshire: with Economical and Political Observations relative to matters applicable to, and illustrated by, the State of those Counties respectively.* London: W. Cobbett.

Coleman, J. 1887. *The cattle sheep and pigs of Britain.* London: The Field.

Crewe, G. 1835. *Extracts from journal of Sir George Crewe recording a visit to the Isle of Portland 1835.* Derbyshire Record Office ms. D2375M 180/24/1–2.

Culley, G. 1807. *Observations on Livestock; Containing Hints for Choosing and Improving the Best Breeds of the Most Useful Kinds of Domestic Animals.* London: Wilkie & Robinson.

Darwin, C. 1858. *The Origin of Species.* London: John Murray.

Darwin, C. 1868. *The Variations of Animals and Plants under Domestication.* 2 vols. London: John Murray.

Davidson, H. R. 1966. *The production and marketing of pigs.* London: Longmans.

Davies, W. 1810. *A general view of the agriculture and domestic economy of North Wales.* London.

Defoe, D. 1724–26. *A Tour through the whole Island of Great Britain, 1724–1726.* Reprint Penguin Classics (abridged) 1986.

Dent, A. 1972. *Donkey: the Story of the Ass from East to West.* London: George G. Harrap.

Dixon, H. H. ('The Druid') 1868. Rise and progress of Shorthorns. *Journal of the Royal Agricultural Society of England* **1**: 317–329.

Elwes, H. J. 1913. *Guide to the primitive breeds of sheep and their crosses.* National Agricultural Centre, Kenilworth: Rare Breeds Survival Trust (reprinted 1983).

Evans, G. E. 1960. *The horse in the furrow.* London: Faber.

Evans, M. 1881. The black cattle of south Wales. *Journal of the Bath and West of England Society* Series 3, **13**: 95–106.

Garrard, G. 1800. *A Description of the Different varieties of Oxen common in the British Isles; Embellished with Engravings; Being an Accompaniment to a Set of Models of the Improved Breeds of cattle upon an Exact Scale from Nature, under the Patronage of the Board of Agriculture.* London: J. Smeeton.

Halford, T. *Quoted by* Wenham, S. M. In: *Short history of the breed.* Wrexham: Kerry Hill Flock Book Society.

Hammond, J. 1922. On the relative growth and development of various breeds and crosses of pigs. *Journal of Agricultural Science, Cambridge* **12**: 387–423.

Hammond, J. 1932. Pigs for pork and pigs for bacon. *Journal of the Royal Agricultural Society of England* **93**: 131–145.

Hammond, J. 1947. The Suffolk ram as a crossing sheep. *Suffolk Sheep Society Official Yearbook* **1947**: 19–28.

Hibbert, S. 1822. *A Description of the Shetland Islands.* Edinburgh.

Hill Farming Research Organisation 1979. *Science and hill farming. Twenty-five years of work at the Hill Farming Research Organisation 1954–1979.* Edinburgh.

Howson, T. A. 1928. *Kerry Hills – and some food for reflection.* Wrexham: Kerry Hill Flock Book Society.

Howson, T. A. 1955. The Radnor sheep. Its history and characteristics. *Radnor Sheep Breeders' Flock Book* **1**: 18–24.

Johnson, S. 1775. In: *Journey to the Western Islands of Scotland and Boswell's journal of a tour to the Hebrides with Samuel Johnson L.L.D.* R. W. Chapman (Ed). Oxford: Oxford University Press paperback, 1970.

Layley, G. W. & Malden, W. J. 1935. *Evolution of the British pig.* London: J. Bolt.

Lisle, E. 1757. *Observations on Husbandry, being a published record of his observations on general agriculture during the period 1710–1722.* London.

Long, W. H. 1969. *A survey of the agriculture of Yorkshire.* Royal Agricultural Society of England County Agricultural Surveys No. 6. London: Royal Agricultural Society of England.

Low, D. 1842. *The Breeds of the Domestic Animals of the British Isles.* Vol. I. *The Horse, and the Ox.* Vol. II. *The Sheep, the Goat, and the Hog.* London: Longman, Orme, Brown, Green & Longmans.

Low, D. 1845. *On the Domesticated Animals of the British Islands: Comprehending the Natural and Economical History of Species and Varieties.* London: Longman, Brown, Green & Longmans.

McDonald, J. 1909. *Stephens' Book of the Farm: Dealing exhaustively with every branch of agriculture.* 5th edn. vol. III – Farm Live Stock, pp. 140–148.

Malthus, T. R. 1798. *An Essay on the Principle of Population as it affects the Future Improvement of Society with Remarks on the Speculation of Mr. Godwin, M. Condorcet and other Writers.* London: J. Johnson in St. Paul's Churchyard.

Marshall, W. 1789. *The rural economy of Gloucestershire; including its Dairy, together with the dairy management of North Wiltshire; and the management of orchards and fruit liquor, in Herefordshire.* 2 vols. Gloucester: R. Raikes for G. Nicoll, London.

Martin, M. 1698. *A late voyage to St. Kilda.* London: Brown & Goodwin.

Milner, F. J. 1983. *George Stubbs on Merseyside. Student notes on the paintings and sculpture in Merseyside County Council Collections.* Liverpool: Merseyside County Council Walker Art Gallery Education Service.

Ministry of Agriculture & Fisheries 1920. *British Breeds of Live Stock.* 3rd edn. London: HMSO.

Moore, H. S. 1980. *Henry Moore's Sheep Sketchbook.* London: Thames & Hudson.

Naismyth, J. 1795. *Observations on the different breeds of sheep and the state of sheep farming in the southern districts of Scotland: being the result of a Tour through these parts made under the direction of the Society for the Improvement of British Wool.* Edinburgh: W. Smellie.

Owen, J. B. 1984. Beef and sheep research and development. In: *ADAS Agricultural Research & Development in Wales 1st Conference, Aberystwyth, April 1984,* pp. 15–22.

Pawson, H. C. 1961. *A survey of the agriculture of Northumberland.* Royal Agricultural Society of England County Agricultural Survey No. 3. London: Royal Agricultural Society of England.

Pegler, H. S. Holmes 1910. *The Book of the Goat.* 4th edn. London: L. Upcott Gill.

Pitt, W. 1794. *General view of the agriculture of the county of Staffordshire.* London.

Punchard, F. 1890. Farming in Devon and Cornwall. *Journal of the Royal Agricultural Society of England* Series 3, **1**: 189–276.

Pusey, P, 1842. On the progress of agricultural knowledge during the last four years. *Journal of the Royal Agricultural Society of England* **3**: 169–217.

Read, C. S. 1854. On the farming of Oxfordshire, *Journal of the Royal Agricultural Society of England* **15**: 189–276.

Richardson 1846. *Domestic pigs.* London: Wm. S. Orr.

Robinson, H. G. 1922. Wensleydale sheep. *Journal of the Royal Agricultural Society of England* **83**: 97–109.

Rothschild, W. 1903. *Quoted by* Chevallier, J. B. (1910). Red Poll cattle. *Journal of the Royal Agricultural Society of England* **71**: 46–56.

Rowe, D. J. 1971. The Culleys, Northumberland farmers, 1767–1813. *Agricultural History Review* **19**: 156–174.

Rowlandson, T. 1849. On the breeds of sheep best adapted to different localities. *Journal of the Royal Agricultural Society of England* **10**: 421–453.

Russell, N. 1986. *Like engend'ring like.* Cambridge: Cambridge University Press.

Ryder, M. L. 1983. *Sheep & Man.* London: Duckworth.

Sidney, S. 1871. *The pig.* London: Geo. Routledge & Sons.

Sinclair, J. (Ed). 1791–1799. *The Statistical Account of Scotland.* Vol. XIX: *Orkney and Shetland.* Reprinted 1978 by E. P. Publishing, Edinburgh.

Spencer, S. 1897. *Pigs. Breeds and management.* London: Vinton.

Stout, A. 1980. *The Old Gloucester. The story of a cattle breed.* Gloucester: Alan Sutton Publishing Ltd.

Stuart, D. 1970. *An illustrated history of belted cattle.* Edinburgh & London: Scottish Academic Press.

Tegetmeier, M. B. O. U. & Sutherland, C. L. 1895. *Horses, Asses, Zebras, Mules, and Mule Breeding.* London: Horace Cox.

Terrington Committee. 1960. *Report of the Committee on the proposed experimental importation of Charollais cattle. Cmnd. 1140.* London: HMSO.

Thompson, M. G. V. 1984. A history of the West Country breeds of sheep. *Ark* **11**: 118–121.

Townsend, H. 1810. *Statistical Survey of the County of Cork.* Dublin.

Trow-Smith, R. 1959. *A history of British livestock husbandry 1700–1900.* London: Routledge & Kegan Paul.

Vancouver, C. 1808. *General view of the agriculture of the County of Devon with observations on the means of its improvement. Drawn up for the consideration of the Board of Agriculture and Internal Improvement.* London: R. Phillips.

Wallace, R. & Watson, J. A. Scott 1923. *Farm Livestock of Great Britain.* 5th edn. Edinburgh: Oliver & Boyd.

Walton, J. R. 1984. The diffusion of the Improved Shorthorn breed of cattle in Britain during the eighteenth and nineteenth centuries. *Transactions of the Institute of British Geographers* New Series, **9**: 22–36.

Watson, J. A. Scott & Hobbs, M. E. 1951. *Great farmers.* London: Faber.

Webster, J. 1794. *General View of the Agriculture of Galloway with observations on the means of its improvement.* London.

White, G. 1789. *The Natural History and Antiquities of Selborne, in the County of Southampton: with Engravings, and an Appendix.* London: B. White & Son.

Whitehead, G. K. 1953. *The ancient white cattle of Britain and their descendants.* London: Faber.

Whitehead, G. K. 1972. *The wild goats of Great Britain and Ireland.* Newton Abbot: David & Charles.

Wilson, J. 1909. *The evolution of British Cattle and the fashioning of breeds.* London: Vinton.

Wilson, J. 1916. Asses and mules. In *Livestock of the Farm.* Ed. C. B. Jones. Vol. VI: 171–176 London: Gresham Publishing Co.

Wiseman, J. 1986. *A history of the British pig.* London: Duckworth.

Youatt, W. 1831. *The Horse.* London: Charles Knight.

Youatt, W. 1835. *Cattle; their Breeds, Management and Diseases.* London: Edward Law.

Youatt, W. 1837. *Sheep, their Breeds, Management and Diseases. To which is added, The Mountain Shepherd's Manual.* London: Robert Baldwin.

Youatt, W. 1846. *The Complete Grazier; or, Farmer's and Cattle Breeder's and Dealer's Assistant.* London: Cradock & Co.

Youatt, W. 1847. *The Pig: a Treatise on the Breeds, Management, Feeding, and Medical Treatment, of Swine; with Directions for Salting Pork, and Curing Bacon and Hams.* London: Cradock & Co.

Young, A. 1771. *The Farmer's Tour through the East of England, being the Register of a journey through various Counties of this Kingdom, to enquire into the State of Agriculture, &c . . . by the author of the Farmer's Letter and the Tours through the North and South of England.* London: W. Strahan, W. Nicoll.

Young, A. 1780. *A tour in Ireland: with general observations on the present state of that kingdom, made in the years 1776, 1777, and 1778, and brought down to the end of 1779.* London: T. Cadell, J. Dodsley.

Young, A. 1794. *General view of the agriculture of the county of Suffolk. Drawn up for the consideration of the Board of Agriculture and Internal Improvement.* London: Sherwood, Neely & Jones.

Young, A. 1804. *General view of the agriculture of the county of Norfolk. Drawn up for the consideration of the Board of Agriculture and Internal Improvement.* London: G. W. Nicoll/Board of Agriculture.

BACKGROUND READING

GENERAL

Bonser, K. J. 1970. *The Drovers.* London: Macmillan.

Bowman, J. C. & Aindow, C. T. 1973. Genetic conservation and the less common breeds of British cattle, pigs and sheep. *University of Reading Department of Agriculture and Horticulture Study* No. 13.

British Wool Marketing Board 1985. *British sheep & wool.* Bradford.

Coppock, J. T. 1971. *An agricultural geography of Great Britain.* London: Bell.

Ernle, Lord 1961. *English farming past and present.* 6th edn., with introductions by G. E. Fussell & O. R. McGregor. London: Heinemann/Frank Cass.

Fussell, G. E. 1966. *The English dairy farmer.* London: Frank Cass.

Grundy, J. 1984. *The origin and development of farm livestock breeds – a reading guide.* National Agricultural Centre, Kenilworth: Rare Breeds Survival Trust.

Housman, W. 1894. Robert Bakewell. *Journal of the Royal Agricultural Society of England* Series 3, **5**: 1–31.

Martin, W. C. L. 1847. *The Ox.* London: Farmer's Library.

Mingay, G. E. 1975. *Arthur Young and his times.* London: Macmillan.

National Sheep Association 1982. *British sheep.* Tring.

Pawson, H. C. 1957. *Robert Bakewell. Pioneer livestock breeder.* London: Crosby Lockwood.

Rowlands, I. W. 1964. Rare breeds of domesticated animals being preserved by the Zoological Society of London. *Nature* **202**: 131–132.

Ryder, M. L. 1964. The history of sheep breeds in Britain. *Agricultural History Review* **12**: 1–12; 65–82.

Simpson, E. S. 1958. The cattle population of England and Wales: its breed structure and distribution. *Geographical Studies* **5**: 45–60.

Urquhart, J. 1983. *Animals on the Farm. Their History from the Earliest Times to the Present Day.* London: MacDonald & Co.

CATTLE

Aberdeen -Angus

Anon. 1978. John C. Graham: Pioneer of bigger Angus. *Aberdeen -Angus Review* No. 60: 80–81.

Barclay, J. R. & Keith, A. 1958. *The Aberdeen -Angus breed – a history.* Perth: Aberdeen-Angus Cattle Society.

Gillanders, E. J. (Ed) 1977. *Aberdeen -Angus official reference book and breed history.* Perth: Holmes McDougall/Aberdeen-Angus Cattle Society.

Hammond, J. 1960. Beef from the dairy herd. *Journal of the British Dairy Farmers' Association* **64**: 13–19.

McCombie, W. 1867. *Cattle and cattle breeders.* 1st edn. Edinburgh: Blackwood.

Merchant, W. J. 1985. The 'bovine revolution' in the north of Scotland. *Aberdeen -Angus Review* No. 68: 57.

Perren, R. 1978. *The meat trade in Britain 1840–1914.* London: Routledge & Kegan Paul.

Ayrshire

Anon. 1948. *Animal Breeding Abstracts* **16**: 191.

Anon. 1986. Report on the Ayrshire breeding scheme programme for the year ended March 1985. *Ayrshire Journal* **58**: 54–56.

Ayrshire Cattle Society 1977. *The Ayrshire Journal: Centenary International Issue* (vol. 49).

McQueen, J. D. H. 1961. Milk surpluses in Scotland. *Scottish Geographical Magazine* **77**: 93–105.

Wiener, G. & Yao, T. S. 1952. Growth of the pedigree Ayrshire cattle population in Great Britain. *Empire Journal of Experimental Agriculture* **20**: 195–208.

Galloway

Biggar, J. 1934. The evolution of British breeds of cattle, with special reference to the Galloways. *Cambridge University Agricultural Society Magazine* **4** (2): 50–52.

Dodd, J. P. 1980. The agriculture of south-western Scotland in the mid-nineteenth century. *Transactions of the Dumfries and Galloway Natural History and Antiquarian Society* **55**: 133–143.

Farmers Weekly 23 May 1986.

Galloway Cattle Society 1919. *Galloway cattle. Their history, characteristics, and value as pure and crossing stock.* Dumfries: Courier & Herald Press.

Galloway Cattle Society 1951. *Galloway cattle.* Dumfries: Courier Press.

Russel, A. J. F. 1981. Beef cattle research. In: *The Hill Farming Research Organisation. Biennial Report 1979–81*, Ed. R. G. Gunn, 145–167. Penicuik: Hill Farming Research Organisation.

Highland

Haldane, A. R. B. 1952. *The drove roads of Scotland.* London: Nelson.

Highland Cattle Society 1984. *Highland Breeders' Journal. 1984 centenary issue.* Thornhill, Dumfries.

Jamieson, A. 1966. The distribution of transferrin genes in cattle. *Heredity* **21**: 191–218.

Shetland

Fenton, A. 1969. Draught oxen in Britain. *Narodopisny Vestnik Ceskoslovensky* **3**: 17–51.

Kerry & Dexter

British Kerry Cattle Society 1964. Kerry cattle. *Journal of the Royal Association of British Dairy Farmers* **68**: 10–13.

Dexter Cattle Society 1964. Dexter cattle. *Journal of the Royal Association of British Dairy Farmers* **68**: 14–19.

Moyles, M. G. 1956. Kerry cattle. A brief outline of the breed's history and development. *Journal of the Department of Agriculture, Dublin* **53**: 53–69.

Thrower, W. R. 1954. *The Dexter cow and cattle keeping on a small scale.* London: Faber.

Young, G. B. 1953. Population dynamics of the Dexter breed of cattle. *Journal of Agricultural Science, Cambridge* **43**: 369–374.

Irish Moiled

Anon. 1984. British cattle. *Stamp collecting* 19 January 1984.

Welsh Black

Colyer, R. J. 1976. *The Welsh cattle drovers.* Cardiff: University of Wales Press.

Edmunds, H. 1981. Welsh Black cattle. *Journal of the Royal Agricultural Society of England* **142**: 24–33.

White Park

Rackham, O. 1986. *The history of the countryside.* London: J. M. Dent.

Chillingham

Bilton, L. 1957. The Chillingham herd of wild cattle. *Transactions of the Natural History Society of Northumberland and Durham* **12**: 137–160.

Hall, S. J. G. 1985. The Chillingham white cattle. *British Cattle Breeders Club Digest* **40**: 24–28.

Hall, S. J. G. 1986. Chillingham cattle: dominance and affinities and access to supplementary food. *Ethology* **71**: 201–215.

Hall, S. J. G. (in press). Chillingham Park and its herd of white cattle: relationships between vegetation classes and patterns of range use. *Journal of Applied Ecology.*

Hall, S. J. G. (in press). Chillingham cattle: social and maintenance behaviour in an ungulate which breeds all year round. *Animal Behaviour.*

Hall, S. J. G. & Hall, J. G. (in press). Inbreeding and population dynamics of the Chillingham cattle (*Bos taurus*). *Journal of Zoology*, London.

Hall, S. J. G., Vince, M. A., Walser, E. S. & Garson, P. J. 1988. Vocalisations of the Chillingham cattle. *Behaviour* **104**: 78–104.

Tankerville, 8th Earl of 1956. The wild white cattle of Chillingham. *Agriculture* **63**: 176–179.

British White

Whitehead, G. K. 1953. *The ancient white cattle of Britain and their descendants.* London: Faber.

Shorthorn

Bowman, J. C. & Hocking, P. M. 1974. The development of a new red and white breed of cattle in the United Kingdom. *Livestock Production Science* **1**: 401–409.

Cadzow, D. J. 1967. The making of a new breed of cattle – the Luing breed. *British Cattle Breeders Club Digest* **23**: 37–44.

Clayton, G. A. 1956. Aspects of breed structure in pedigree British Shorthorn cattle. *Proceedings of the British Society of Animal Production* **1956**: 107.

Hocking, P. M. 1979. The Shorthorn Society's experimental breeding programme: results and plans for the future. *British Cattle Breeders Club Digest* **34**: 40–46.

Luing Cattle Society. 1966. *What is this 'Luing' breed?* Pencaitland.

McDougall, J. D. 1978. Blue-Greys as suckler cows and the role of the Whitebred Shorthorn. *ADAS Quarterly Review* No. 30: 153–158.

Mason, I. L. 1969. *A world dictionary of livestock breeds types and varieties.* Farnham Royal: Commonwealth Agricultural Bureaux.

Marson, T. B. 1946. *The Scotch Shorthorn.* Edinburgh: The Scottish Shorthorn Breeders' Association.

Mingay, G. 1982. *British Friesians: an epic of progress.* Rickmansworth: British Friesian Cattle Society.

Robinson, H. G. 1928. The Cumberland and Westmorland Shorthorn. *Journal of the Royal Agricultural Society of England* **89**: 109–125.

Walton, J. R. 1984. The diffusion of the Improved Shorthorn breed of cattle in Britain during the eighteenth and nineteenth centuries. *Transactions of the Institution of British Geographers* New Series, **9**: 22–36.

Wright, S. 1923. Mendelian analysis of the pure breeds of livestock. II. The Duchess family of Shorthorns as bred by Thomas Bates. *Journal of Heredity* **14**: 339–348.

Blue Albion

Cheese, A. 1980. *Blue Albion Cattle. Shugborough Information Sheet.* Stafford: Staffordshire County Museum Service.

Lincoln Red

Edgar, C. D. 1983. E. L. C. Pentecost – a solitary cattle breeder. *Journal of the Royal Agricultural Society of England* **144**: 61–73.

Grigg, D. 1966. *Agricultural Revolution in South Lincolnshire.* Cambridge University Press.

Longhorn

Clutton-Brock, J. 1982. British cattle in the 18th century. *Ark* **9**: 55–59.

Pawson, H. C. 1957. *Robert Bakewell. Pioneer livestock breeder.* London: Crosby, Lockwood & Son.

Russell, N. C. 1981. Who improved the eighteenth-century Longhorn cow? *Exeter Papers in Economic History* **14**: 199–240.

Devon

Fussell, G. E. 1949. Francis Quartly and Devon cattle. *Estate Magazine* **49**: 83–85.

South Devon

Anon. 1969. *Animal Breeding Abstracts* **37**: 168.

Baker, C. M. A. 1984. The origin of South Devon cattle. *Agricultural History Review* **32**: 145–158.

Bangham, A. D. 1963. Cattle haemoglobins. *In:* Man & Cattle (Ed. by A. E. Mourant & F. E. Zeuner). *Royal Anthropological Institute Occasional Paper* No. 18: 34–40.

Benyon, V. G. 1975. The changing structure of dairying in Devon. *Devon Historian* **11**: 35–40.

Devon Farming Study Group 1952. *Devon Farming. A first study incorporating a review and a cartographic and written analysis of cropping and stocking.*

Johnson, R. F. 1951. South Devons for profitable beef. *Farming in Gloucestershire* **10**: 49–50.

McKellar, J. C. & Ouhayoun, J. 1973. Studies on the double-muscle character. VIII – Frequency and expression in South Devon cattle. *Annales de génétique et de sélection animale* **5**: 163–176.

Turpitt, W. G. 1954. The cattle of Britain. 18. South Devon. *Agriculture* **61**: 446–447.

Gloucester

Gethyn-Jones, J. E. 1986. *The Jenner Museum, Berkeley, Gloucestershire.* Berkeley: Jenner Trust.

Hereford

Hammond, J. 1960. Beef from the dairy herd. *Journal of the Royal Association of British Dairy Farmers* **64:** 13–19.

Heath-Agnew, E. 1983. *A history of Hereford Cattle and their breeders.* London: Duckworth.

Jones, E. L. 1964. Hereford cattle and Ryeland sheep: economic aspects of breed changes, 1780–1870. *Transactions of the Woolhope Naturalists' Field Club* **38:** 36–48.

Southgate, J. R. 1981. The current practice of commercial cross-breeding in the UK with particular reference to the effects of breed choice. In: *Beef production from different dairy breeds and dairy beef crosses* (Ed. G. J. More O'Ferrall). The Hague/CEC: Martinus Nijhoff.

Sussex

Law, W. 1924. Our windmills. *Brighton & Hove Archaeologist* No. 2: 87–97.

Red Poll

Cockburn, E. O. 1962. John Hammond – the man. *Animal Production* **4:** 1–12.

Donald, H. P. 1945. The growth and distribution of the pedigree Red Poll cattle population in England. *Empire Journal of Experimental Agriculture* **13:** 169–183.

Red Poll Cattle Society 1933. *Reprint of Red Polled Herd Book Index Volume (of Vols. 1 to 6) containing History of the Breed, Groups and Tribes.* Ipswich.

Trist, P. J. O. 1971. A survey of the agriculture of Suffolk. *Royal Agricultural Society of England County Agricultural Surveys* No. 7. London: Royal Agricultural Society of England.

Black and white cattle

Anon. 1954. *Animal Breeding Abstracts* **22:** 172.

Anon. 1956. *Animal Breeding Abstracts* **24:** 84.

Farmers Weekly 23 May 1986.

Milk Producer March 1984 p. 10.

Mingay, G. 1982. *British Friesians: an epic of progress.* Rickmansworth: British Friesian Cattle Society.

Channel Island cattle

Bell, R. W. 1979. *The history of the Jersey Cattle Society of the United Kingdom, 1878–1978.* Reading: Jersey Cattle Society.

Butler-Hogg, B. W. & Wood, J. D. 1982. The partition of body fat in British Friesian and Jersey steers. *Animal Production* **35:** 253–262.

Miller, D. 1986. "We agriculturists of England". Farm animals of Queen Victoria and Prince Albert. *Country Life* June 1986: 1824–1826.

Continental beef breeds

Anon. 1961. *Animal Breeding Abstracts* **29:** 376.

Anon. 1970. *Animal Breeding Abstracts* **38:** 707.

Edwards, J. et al. 1966. *The Charolais report.* Thames Ditton: Milk Marketing Board.

Holmes, W. 1976 (Chmn.) *Report of the evaluation of the first importation into Great Britain in 1970–71 of Limousin bulls from France and Simmental bulls from Germany and Switzerland.* London: HMSO.

Holmes, W. 1984. Technical and economic background to beef production in Britain. In: *Grassland Beef Production Conference Proceedings Wye College, July 25–27, 1983.*

Kempster, A. J. & Southgate, J. R. 1984. Beef breed comparisons in the U.K. *Livestock Production Science* **11:** 491–501.

SHEEP

Parkland and primitive sheep

Aberdeen School of Agriculture 1983. Comparison of Cheviot × Shetland cross ewes with Greyface ewes under intensive systems with housing and shearing. *Research Investigations and Field Trials* **1983:** 35–37.

Allan, R. J. P. 1984. Scottish island sheep. *Ark* **11:** 162–166.

Bowie, H. M. S. & Bowie, S. H. U. 1983. Shetland sheep. *Ark* **10:** 128–129.

Bowie, S. H. U. 1987. Shetland's native farm animals. Part Two – Shetland Sheep. *Ark* **14:** 194–199.

Farmers Weekly 22 March 1957.

Hall, S. J. G. 1975. Some recent observations on Orkney sheep. *Mammal Review* **5:** 59–64.

Hall, S. J. G. 1986. Genetic conservation of rare British sheep: the Portland, Manx Loghtan and Hebridean breeds. *Journal of Agricultural Science*, Cambridge **107:** 133–144.

Jewell, P. A. 1980. The Soay sheep – Parts 1 and 2. *Ark* **7:** 51–57; 87–93.

Jewell, P. A. 1986. Survival in a feral population of primitive sheep on St. Kilda, Outer Hebrides, Scotland. *National Geographic Research* **2:** 402–406.

Jewell, P. A., Milner, C. & Boyd, J. Morton 1974. *Island survivors: the ecology of the Soay sheep of St. Kilda.* London: University of London Athlone Press.

Lockley, R. M. 1960. Wild sheep in Wales. *Nature in Wales* **6:** 75–78.

Paterson, I. W. 1984. The foraging strategy of the seaweed-eating sheep on North Ronaldsay, Orkney. PhD thesis, University of Cambridge.

Ryder, M. L. 1982. Shetland sheep and wool. *Ark* **9:** 93–102.

Sheep of hill and mountain

Al-Nakib, F. M. S., Findlay, R. H. & Smith, C. 1986. Performance of different Scottish Blackface stocks and their crosses. *Journal of Agricultural Science*, Cambridge **107:** 119–123.

Anderson, A.. W. 1928. The live stock of Yorkshire. *Journal of the Royal Agricultural Society of England* **89:** 67–77.

Anon. 1971. *Animal Breeding Abstracts* **39:** 806.

Anon. 1974. The early history of the Blackface Sheep Breeders' Association. *Blackface Journal* **26:** 32.

Barrington, J. 1984. *Red sky at night.* London: Michael Joseph.

Boyd, J. Morton 1981. The Boreray sheep of St. Kilda, Outer Hebrides, Scotland: the natural history of a feral population. *Biological Conservation* **20:** 215–227.

Bullock, D. J. 1983. Borerays, the other rare breed on St. Kilda. *Ark* **10:** 274–278.

Carlyle, W. J. 1978. The distribution of store sheep from markets in Scotland. *Transactions of the Institution of British Geographers* New Series **3:** 226–245.

Carlyle, W. J. 1979. The changing distribution of breeds of sheep in Scotland, 1795–1965. *Agricultural History Review* **27:** 19–29.

Dent, L. Y. & Cooper, M. McG. 1956. Sheep crosses in north-east England. *Agriculture* **63**: 173–175.

Evans, W. M. R. 1958. The Welsh half-bred ewe. *Agriculture* **65**: 509–512.

Evans, W. M. R. & Vale, T. W. 1953. The Welsh half-bred ewe. *Journal of the Royal Welsh Agricultural Society* **1953**: 55.

Firbank, T. 1940. *I bought a mountain*. London: Harrap.

Halsall, E. 1985. *Derbyshire Gritstone*. Cliviger, Burnley: Derbyshire Gritstone Sheepbreeders' Society.

Kippax, S. E. 1953. Sheep on the northern hills. *Agriculture* **60**: 168–172.

Kirby, H. R. 1954. Some examples of breed changes in sheep. *N.A.A.S. Quarterly Review*, No. 23: 136.

Kirby, H. R. 1954. The Yorkshire Pennines and their sheep. *Journal of the Yorkshire Agricultural Society* No. 105: 35–42.

Owen, J. B. 1981. Hill sheep production. In: *The effective use of forage and animal resources in the hills and uplands*. Occasional Symposium No. 12, p.59. British Grassland Society.

Rae, A. L. 1952. Crossbreeding of sheep. *Animal Breeding Abstracts* **20**: 197–207.

Roberton, R. J. 1953. Border sheep. *Scottish Agriculture* **32**: 12–15.

Steane, D. E. 1983. The significance of interactions in practical sheep breeding in northern Europe. *Livestock Production Science* **10**: 39–48.

Stewart, W. L. 1951. British breeds of livestock. V. The hill sheep of the north. *British Agricultural Bulletin* **3**: 216–221.

Wilkinson, J. H. S. 1949. Welsh Mountain ewes as a source of grassland breeding ewes. *Journal of the Royal Agricultural Society of England* **110**: 76–88.

Long-woolled breeds of sheep

Allanson, G. 1961. *Kent or Romney Marsh sheep. A study of a longwool breed in competition*. Ashford: Wye College, University of London.

Armitage, P. L. 1983. The early history of English longwool sheep. *Ark* **10**: 90–97.

Carter, E. S. 1970. The agriculture of Lincolnshire. *Journal of the Royal Agricultural Society of England* **131**: 56–68.

Devon Farming Study Group 1952. *Devon Farming. A first study incorporating a review and a cartographic and written analysis of cropping and stocking*.

Farmers Weekly 23 May 1986 p.35.

Filmer, R. 1980. The Kent sheep: an ancient breed. *Bygone Kent* **1**: 387–395.

Perkins, J. A. 1977. *Sheep farming in eighteenth and nineteenth century Lincolnshire*. Grimsby: Society for Lincolnshire History and Archaeology Occasional Paper No. 4.

Wiener, G. 1961. Population dynamics in fourteen lowland breeds of sheep in Great Britain. *Journal of Agricultural Science*, Cambridge **57**: 21–28.

The Merino experiment

Carter, H. B. 1964. *His Majesty's Spanish Flock. Sir Joseph Banks and the Merinos of George III of England*. London: Angus & Robertson.

New Leicester and the crossing sires; the rising stars

Anderson, A. W. 1928. The live stock of Yorkshire. *Journal of the Royal Agricultural Society of England* **89**: 67–77.

Anon. 1951. The Texel sheep. *Agriculture* **58**: 398.

Anon. 1975. *Animal Breeding Abstracts* **43**: 321.

Bosanquet, C. I. C. 1974. The story of a Border Leicester flock. *Journal of the University of Newcastle Agricultural Society* **25**: 30–32.

Burns, M. 1967. The Katsina wool project II – coat and skin data from $\frac{3}{4}$-Merino and Wensleydale crosses. *Tropical Agriculture*, Trinidad **44**: 253–274.

Hunt, J. 1986. *The Bluefaced Leicester. A history of the breed*. Preston: Bluefaced Leicester Sheep Breeders' Association.

Macdonald, S. 1974. The role of George Culley of Fenton in the development of Northumberland agriculture. *Archaeologia Aeliana* 5th Series, **3**: 131–141.

Robinson, H. G. 1922. Wensleydale sheep. *Journal of the Royal Agricultural Society of England* **83**: 97–109.

Smith, C. 1982. ABRO experiments with crossing-ram breeds. In: *Animal Breeding Research Organisation report 1982*. Edinburgh: Animal Breeding Research Organisation.

Speedy, A. W. 1980. *Sheep production: science into practice*. London: Longmans.

Williams, J. C., Crees, S. & Owen, J. B. 1984. The Cambridge sheep project. In *ADAS Agricultural Research & Development in Wales 1st Conference, Aberystwyth, April 1984*, p.43.

Wolf, B. T. 1980. The choice of terminal sire for lamb carcass production. *Journal of the Agricultural Society of University College Wales* **61**: 106–119.

Young, G. B. & Purser, A. F. 1962. Breed structure and genetic analysis of Border Leicester sheep. *Animal Production* **4**: 379–389.

Upland sheep

Colyer, R. J. 1983. Aspects of the pastoral economy in pre-industrial Wales. *Journal of the Royal Agricultural Society of England* **144**: 45–60.

Evans, W. 1951. An ecological survey of hill sheep farming in Breconshire. *Agriculture* **58**: 5–13.

Hafez, E. S. E. 1952. Studies on the breeding season and reproduction of the ewe. Parts I-III. *Journal of Agricultural Science*, Cambridge **42**: 189–265.

Hall, S. J. G. 1986. Genetic conservation of rare British sheep: the Portland, Manx Loghtan and Hebridean breeds. *Journal of Agricultural Science*, Cambridge **107**: 133–144.

Hill, R. 1984. *Shropshire sheep. A history*. Shrewsbury: Shropshire County Museum Service.

Jones, E. L. 1964. Hereford cattle and Ryeland sheep: economic aspects of breed changes, 1780–1870. *Transactions of the Woolhope Naturalists' Field Club* **38**: 36–48.

Robinson, J. F. 1956. Sheep on the English-Welsh border. *Agriculture* **56**: 178–183.

Stevens, M. R. A. 1948. A geographical survey of the sheep-rearing industry of Knighton and neighbourhood. *Transactions of the Radnorshire Society* **18**: 38–45.

Wheeler, A. G. & Land, R. B. 1973. Breed and seasonal variation in the incidence of oestrus and ovulation in the sheep. *Journal of Reproduction and Fertility* **35**: 583–584.

Williams, S. M. 1954. Fertility in Clun Forest sheep. *Journal of Agricultural Science*, Cambridge **45**: 202–228.

Downland sheep

Bowie, G. 1984. Sheep husbandry in Hampshire in the nineteenth century. *Hampshire County Magazine* Sept 1984: 61–63; Oct. 1984: 45–46.

Farrant, S. 1978. John Ellman of Glynde in Sussex. *Agricultural History Review* **26**: 77–88.

French, S. 1955. Jonas Webb of Babraham. *East Anglian Magazine* **14:** 275–278.

Jesse, R. H. B. 1960. *A survey of the agriculture of Sussex. Royal Agricultural Society of England County Agricultural Surveys* No. 2. London: Royal Agricultural Society of England.

Jones, E. L. 1961. The entry of Southdown sheep into the Wessex chalklands. *The Laden Wain, Journal of the Agricultural Club, University of Reading* **1961:** 38–40.

Kirk, J. (unpubl. ms). History of the Dorset Horn. Dorchester: Dorset Horn Sheep Breeders' Association.

Lloyd, E. W. 1928. John Ellman of Glynde: his life-work and correspondence. *Journal of the Royal Agricultural Society of England* **89:** 32–50.

Lloyd, E. W. (Ed). 1936. *The Southdown sheep.* 3rd edn. Chichester: W. G. & T. R. Willis.

Ministry of Agriculture, Fisheries & Food 1982. *Breeding lambs for the market – choosing a terminal sire breed. MAFF Leaflet 822.* Alnwick: Ministry of Agriculture, Fisheries & Food.

Rayns, F. 1969. Norfolk Horn sheep. *Journal of the Royal Agricultural Society of England* **130:** 20–30.

Trist, P. J. O. 1971. A survey of the agriculture of Suffolk. *Royal Agricultural Society of England County Agricultural Surveys* No. 7. London: Royal Agricultural Society of England.

Walton, J. R. 1983. The diffusion of improved sheep breeds in eighteenth- and nineteenth-century Oxfordshire. *Journal of Historical Geography* **9:** 175–195.

GOATS

British Goat Society *Breeds of goats and breed points.* Bury St. Edmunds: British Goat Society.

Carruthers, S. P. (Ed). 1986. *Alternative enterprises for agriculture in the UK. CAS Report 11.* Reading: Centre for Agricultural Strategy.

PIGS

Foster, J. 1983. Gloucester Old Spots going strong. *Ark* **10:** 83–85.

Hall, S. J. G. (in press). Breed structures of rare pigs: implications for genetic conservation of the Berkshire, Tamworth, Middle White, Large Black, Gloucester Old Spot, British Saddleback and British Lop. *Conservation Biology.*

Hunt, R. 1983. Farewell to the Lincolnshire Curly Coat. *Ark* **10:** 119–121.

Ministry of Agriculture, Fisheries & Food 1983. *Pigs: the outdoor breeding herd. MAFF Booklet 2431.* Alnwick: Ministry of Agriculture, Fisheries & Food.

Meat and Livestock Commission 1986. *Pig Yearbook 1986.* Bletchley.

Steane, D. E. & Cloke, G. J. 1983. The Chester White in the USA and in Britain. *Ark* **10:** 122–125.

HORSES, PONIES, AND THE DONKEY

Arbuthnott, Lord 1965. *The Rhum stud of Highland ponies.* Edinburgh: Nature Conservancy.

Askew, R. P. 1937. The future changes in the numbers of horses in England and Wales. *Farm Economist* **2:** 129–133.

Baird, E. 1982. *The Clydesdale horse.* London: Batsford.

Britton, D. K. 1960. The disappearance of the farm horse in England and Wales. *Incorporated Statistician* **10:** 79–88.

Calder, A. 1927. The role of inbreeding in the development of the Clydesdale breed of horses. *Proceedings of the Royal Society of Edinburgh* **47:** 118.

Chivers, K. 1976. *The Shire horse: a history of the breed, the Society and the men.* London: J. A. Allen.

Chivers, K. 1988. *History with a Future. Harnessing the heavy horse for the 21st century.* Peterborough: Shire Horse Society/Royal Agricultural Society of England.

Clutton-Brock, T. H. (Ed). 1986. *Rhum: the natural history of an island.* Edinburgh: Edinburgh University Press.

Hulme, S. 1980. *Native ponies of the British Isles.* Hindhead, Surrey: Saiga Publishing.

Pennell, N. 1959. The story of 'Dyoll Starlight' 1894–1929. *Journal of the Royal Welsh Agricultural Society* **28:** 31–38.

Putman, R. J. 1986. *Grazing in temperate ecosystems: large herbivores and the ecology of the New Forest.* London: Croom Helm.

Thompson, F. M. L. 1970. Victorian England: the horse-drawn society. An inaugural lecture, 22 October 1970, Bedford College, University of London.

Thompson, F. M. L. (Ed). 1983. *Horses in European Economic History: a Preliminary Canter.* Reading: British Agricultural History Society.

LIVESTOCK ART

Anon. 1966. *Animal painting. Van Dyke to Nolan. The Queen's Gallery, Buckingham Palace.* London: privately published.

Anon. 1985. Obituary: Mr John Gilroy. *The Times,* 13 April 1985.

Bateman, J. 1957. *Oil painting.* London: The Studio Publications.

Boalch, D. H. 1958. *Prints and paintings of British farm livestock 1780–1910.* Harpenden: Rothamsted Experimental Station Library. Library.

Clutton-Brock, J. 1976. The models of livestock made by George Garrard (1760–1826) that are in the British Museum (Natural History). *Agricultural History Review* **24:** 18–29.

Clutton-Brock, J. 1979. George Garrard's models of sheep. *Textile History* **10:** 203–206.

Niall, I. 1985. *Portrait of a country artist. C. F. Tunnicliffe RA 1901–1979.* London: Gollancz.

Satterley, G. 1983. *Life in Caithness and Sutherland. Photographs by Glyn Satterley with an Introduction by Bette McArdle.* Edinburgh: Paul Harris Publishing.

Picture credits

Front cover   *On a farm in East Kent* by T. S. Cooper RA. By courtesy of Mr David Messum.

1   Introduction

Page
2   The Pass of Leny by Gourlay Steell. Reproduced by gracious permission of Her Majesty The Queen.
10   Hereford cow and calf. By courtesy of Diane Rosher.
13   Eclipse by Francis Sartorius. By courtesy of The Jockey Club.
15   Robert Bakewell by Boultbee. By courtesy of Leicestershire Museums and Art Galleries, Leicester.

2   Western and northern cattle

18   Aberdeen-Angus cow. By courtesy of the Aberdeen-Angus Cattle Society.
21   Pirate of Monkwood by Eirene Hunter. By courtesy of the Royal Highland and Agricultural Society of Scotland.
21   Guisachan herd by Gourlay Steell RSA. By courtesy of the Royal Highland and Agricultural Society of Scotland.
23   Ayrshire cows by James Howe. By courtesy of the Trustees of the National Library of Scotland.
24   Fifeshire cow by William Shiels. By courtesy of the National Gallery of Scotland.
25   Ayrshires at the East of England Show. Photographed by Stephen J. G. Hall.
27   Bolebec Dun Champion by Diane Rosher. By courtesy of Mr F. N. Tucker.
28   Galloway bull by James Howe. By courtesy of the Trustees of the National Library of Scotland.
29   Highland cattle by James Howe. By courtesy of the National Gallery of Scotland.
31   Presentation of trophies at the Royal Highland Show. By courtesy of Susie Whitcombe.
32   Shetland draught team by Hibbert. Reproduced by permission of the Syndics of Cambridge University Library.
35   Dexter cattle. Photographed by Stephen J. G. Hall.
38   Stamp picturing Irish Moiled cow. Reproduced by permission of the National Postal Museum.
39   Welsh Black cattle in landscape by Walter Bayes. By courtesy of the Tate Gallery Publications.
40   Benthal and Bran by Daniel Clowes. By courtesy of the Grosvenor Museum, Chester, © Mr G. Edington.
42   White cattle in front of Streatlam Castle by Joseph Miller. By courtesy of The Bowes Museum, Barnard Castle, Co. Durham.
43   Chartley Michael. Photographed by Stephen J. G. Hall.
44   Cow from Haverfordwest by William Shiels. By courtesy of the Royal (Dick) Veterinary College, Edinburgh.
45   Chillingham herd. Photographed by Mr R. C. Pietersma.
46   Chillingham cattle by Edwin Landseer. From the collection of the Laing Art Gallery, Newcastle: reproduced by permission of Tyne and Wear Museums Service.
47   British White bull by W. A. Clark. By courtesy of Mr R. C. Topham.

3   Short-horned cattle

51   The ox by Cuit. By courtesy of Rothamsted Experimental Station.
53   Winbrook Atom 2nd. By courtesy of the Rare Breeds Survival Trust.
55   Shorthorn bull by James Bateman. By courtesy of the Tate Gallery Publications.
56   Shorthorn cow with calf by James Macleod. By courtesy of Mr A. Stirling of Keir.
57   Ryleys herd. Photographed by Stephen J. G. Hall.
58   Lincoln Red Cattle Society Field Day. By courtesy of Mr W. Richardson.
59   Red heifer by J. Tennant. By permission of Sotheby's.

4   English lowland cattle

60   Androsspoll. By courtesy of Susan Crawford.
68   Francis Quartly by Thomas Mogford. By courtesy of Mrs M. Turner.
69   'Coke of Norfolk' by W. H. Davis. By courtesy of the University of Oxford Institute of Agricultural Economics.
69   *The Bail Bull* by Evelyn Dunbar. By courtesy of the Tate Gallery Publications.
71   South Devon bullock by Diane Rosher. By courtesy of Mr W. H. Overton.
73   Wick Court herd. By courtesy of The Honourable Flora Stuart.
75   Blossom by Stephen Jenner. By courtesy of the Jenner Museum.
78   Sale ring at Stocktonbury. By courtesy of the Hereford Herd Book Society, © E. Heath-Agnew Esq.
79   Androsspoll. By courtesy of Susan Crawford.
81   Dyke Road Mill and Sussex oxen. By courtesy of the Royal Pavilion, Art Gallery and Museums (Preston Manor) Brighton.
82   Petworth General 26th by Diane Rosher. By courtesy of Mr P. W. Kelmsley.
84   Starling. By courtesy of the Marquess Townshend.
84   Cow resembling modern Red Poll. By courtesy of the Red Poll Cattle Society.
85   Colmans trademark. By courtesy of Colmans of Norwich.

5   Black and white cattle

86   *Commotion in the Cattle Ring* by James Bateman. By courtesy of the Tate Gallery Publications.
88   Moneymore Bunty 6 by Kenneth Ogborne. By courtesy of the British Friesian Cattle Society.
89   British Friesian heifers. Photographed by Stephen J. G. Hall.

6   Channel Island cattle

90   The Victoria cow by T. S. Cooper. Reproduced by gracious permission of Her Majesty The Queen.

184 Royal Shrewsbury 2nd. By courtesy of the Shropshire Sheep-breeders Association and Flock Book Society.
184 1982 Shropshire Flock Book Centenary. By courtesy of the Shropshire Sheepbreeders Association and Flock Book Society.
185 Witney Bank £5 note. By courtesy of Steve Baylis.
186 John Tredwell's flock by Whitford. By courtesy of Buckinghamshire County Museum.
188 Norfolk Horn sheep. Photographed by Sally Anne Thompson.
189 Bismark. By courtesy of the Suffolk Sheep Society.
190 Suffolk-type sheep. By courtesy of Sylvia Frattini.
191 Three ewes in a landscape by Ralph Whitford. By courtesy of Iona Antiques.
192 Dorset Down sheep in early spring and at shearing time by
& 193 William Gunning. By courtesy of Bonhams.

16 Goats

194 Welsh goats by Shiels. By courtesy of the National Museum of Wales (Welsh Folk Museum).
197 Angora goats. Photographed by Glyn Satterley.
199 Feral goats near Lynton, Exmoor. Photographed by Glyn Satterley.
201 Goat's head from the tomb of Richard Bagot. By courtesy of Revd R. Vaughan, Vicar of Blithfield, Staffordshire: photographed by John Downs, BM(NH) Photographic Unit.
201 Bagot nanny Bemborough Ann. By courtesy of Mr Joe Henson.

17 Pigs

202 'Spherical' pig. By courtesy of Mr R. Dyott.
204 Girl with pigs by George Morland. By courtesy of the University of Oxford Institute of Agricultural Economics.
205 Pigs and their antics. By courtesy of Mrs S. C. Holmes.
206 Heavy Sow by Ted Roocroft. By courtesy of Mr Ted Roocroft: from the collection of Don McKinley.
213 Large White pigs by Sir Stanley Spencer RA. By courtesy of the Tate Gallery Publications.
214 Suffolk and Bedfordshire pigs by J. W. Giles. By courtesy of Rothamsted Experimental Station.
215 Yorkshire hog. By courtesy of Rothamsted Experimental Station.
217 Tamworth sow. Photographed by Sally Anne Thompson.
219 Gloucester Old Spot pig. By courtesy of James Lynch.
221 Ashdonian Imposing 76th by Helen Stevens. By courtesy of Mr F. Ketteridge.
223 Large Black pigs at Litlington, Hertfordshire. Photographed by Keith Huggett.

18 Heavy draught horses

224 Whitbread horse by Garrard. By courtesy of Whitbread and Co.
227 Young & Co's dray by Nina Colmore. By courtesy of Parnassus Gallery.
228 Haymakers by George Stubbs. By courtesy of Lady Lever Art Gallery, Liverpool.
229 Scotch Farm Horse by James Howe. By courtesy of the Trustees of the National Library of Scotland.
230 Model of Clydesdales at work. By courtesy of Judy Boyt.
230 Peeblesshire fair by James Howe. By courtesy of the Scottish National Portrait Gallery.
231 Bonnie Buchlyvie. By courtesy of Beamish, North of England Open Air Museum, Beamish, Co. Durham.
231 Trace Horses on the Turn by James Howe. By courtesy of East Lothian District Libraries.
233 Mares and foals near Ipswich by Sidney Seymour Lucas. By courtesy of Ipswich Museums and Galleries.
234 Sudbourne Premier by Herbert Haseltine. By courtesy of the Tate Gallery Publications.
236 Pony on Dartmoor. Photographed by Glyn Satterley.

19 Light horses, ponies, and donkeys

238 Masterful Jack. Photographed by Stephen J. G. Hall.
240 The Vardy Horse. By courtesy of Beamish, North of England Open Air Museum, Beamish, Co. Durham.
241 Fell pony by Taylor Longmire. By courtesy of The National Trust.
242–3 'Colt-hunt' by Lucy Kemp-Welch. By courtesy of the Tate Gallery Publications.
244 The Dales pony. By courtesy of Beamish, North of England Open Air Museum, Beamish, Co. Durham.
245 Exmoor ponies. By courtesy of Mrs A. Dent.
246 Welsh pony. Photographed by Sally Anne Thompson.
247 Old Douse and Bryn Arth Madonna. By courtesy of Dr Wynne Davies.
248 Dyoll Starlight. By courtesy of Dr Wynne Davies.
249 Hebridean ponies by Shiels. By courtesy of the National Museum of Antiquities of Scotland.
250 Highland ponies by Susan Crawford. By courtesy of Mr Duncan Cameron of Lochliel.
253 The Pallisson children and their pet donkey by Thomas Lawrence. By courtesy of the National Trust, Polesden Lacey, Surrey.
254 Lady with her helper. By courtesy of Dr A. M. Hall.
254 Donkey carrying peat. By courtesy of the Ulster Folk Museum.
255 Mule train. By courtesy of the National Museum of Wales.

Back cover Swaledale sheep in the Yorkshire Dales. By courtesy of Glyn Satterley.

We would particularly like to thank John Downs of the British Museum (Natural History) Photographic Unit for photographing the following pictures: British White bull (p.47), Francis Quartly (p.68), 'Coke of Norfolk' (p.69), Blossom (p.74), Dyke Road Mill (p.80), Starling (p.84), flock of Cotswolds (p.143), young Reggie Benham (p.168), pigs and their antics (p.204), 'Spherical pig' (p.203), Berkshires (p.208), Ashdonian Imposing 76th (p.221), and Highland ponies (p.250).

All Tunnicliffe drawings are reproduced from Portrait of a country artist by Ian Niall, published by Thames & Hudson.

Index